Connecting Families

Carman Neustaedter · Steve Harrison
Abigail Sellen
Editors

Connecting Families

The Impact of New Communication
Technologies on Domestic Life

 Springer

Editors
Carman Neustaedter
School of Interactive Arts
   and Technology
Simon Fraser University
Surrey, BC, Canada

Abigail Sellen
Socio-Digital Systems
Microsoft Research Cambridge
Cambridge, UK

Steve Harrison
Department of Computer Science
   and School of Visual Arts
Virginia Polytechnic Institute
   and State University
Blacksburg, VA, USA

ISBN 978-1-4471-4191-4   ISBN 978-1-4471-4192-1 (eBook)
DOI 10.1007/978-1-4471-4192-1
Springer London Dordrecht Heidelberg New York

Library of Congress Control Number: 2012943353

Printed on acid-free paper

Springer is part of Springer Science + Business Media (www.springer.com)

Contents

Contributors

Gregory D. Abowd Georgia Institute of Technology, School of Interactive Computing, GVU Center, Atlanta, GA, USA
e-mail: abowd@gatech.edu

Ronald M. Baecker Technologies for Aging Gracefully Lab, University of Toronto, Toronto, Canada
e-mail: ron@taglab.ca

Rafael Ballagas Nokia Research Center, IDEA Group, 200 South Mathilda Ave., Sunnyvale, CA 94086, USA
e-mail: tico.ballagas@nokia.com

Stacy Branham Department of Computer Science and School of Visual Arts, Center for Human-Computer Interaction, Virginia Polytechnic Institute and State University, 2202 Kraft Drive, Blacksburg, VA 24060, USA
e-mail: sbranham@vt.edu

Xiang Cao Microsoft Research Asia, Beijing, China
e-mail: xiangc@microsoft.com

Jessica David Technologies for Aging Gracefully Lab, University of Toronto, Toronto, Canada
e-mail: jessicam.david@utoronto.ca

Saul Greenberg Department of Computer Science, University of Calgary, 2500 University Drive NW, Calgary, AB T2N 1N4, Canada
e-mail: saul.greenberg@ucalgary.ca

Steve Harrison Department of Computer Science and School of Visual Arts, Center for Human-Computer Interaction, Virginia Polytechnic Institute and State University, 2202 Kraft Drive, Blacksburg, VA 24060, USA
e-mail: srh@vt.edu

Kori M. Inkpen Microsoft Research, 1 Microsoft Way, Redmond, WA 98052, USA
e-mail: kori@microsoft.com

Tejinder K. Judge   Google Inc., Mountain View, CA, USA
e-mail: tkjudge@google.com

Joseph 'Jofish' Kaye   Nokia Research Center, IDEA Group, 200 South Mathilda
Ave., Sunnyvale, CA 94086, USA
e-mail: jofish.kaye@nokia.com

Panos Markopoulos   Eindhoven University of Technology,
Eindhoven, Netherlands
e-mail: p.markopoulos@tue.nl

Karyn Moffatt   School of Information Studies, McGill University,
3661 Peel Street, Rm 303C, Montreal, QC, H3A 1X1, Canada
e-mail: karyn.moffatt@mcgill.ca

Carman Neustaedter   School of Interactive Arts and Technology, Simon Fraser
University, 102 Avenue, 250-13450, V3T 0A3 Surrey, BC, Canada
e-mail: carman_neustaedter@sfu.ca

Hayes Raffle   Staff Interaction Designer, Google, 1600 Amphitheater Parkway,
Mountain View, CA 94043, USA
e-mail: hraffle@google.com

Ruud Schatorjé   Eindhoven University of Technology, Eindhoven, Netherlands
e-mail: mail@ruudschatorje.nl

Abigail Sellen   Socio-Digital Systems, Microsoft Research Cambridge,
JJ Thomson Avenue 7, Cambridge CB3 0FB, UK
e-mail: asellen@microsoft.com

Svetlana Yarosh   Georgia Institute of Technology, School of Interactive
Computing, GVU Center, Atlanta, GA, USA
e-mail: lana@cc.gatech.edu

Chapter 1
Connecting Families: An Introduction

Carman Neustaedter, Steve Harrison and Abigail Sellen

Abstract Family life is complex and dynamic. It forms a core part of our existence. Underpinning family life, is *family connection*: how families not just communicate with each other, but how they share their lives and routines, how they engage in social touch, and how they negotiate being together, or being apart. This book explores the various ways in which family members "connect" within the same household, across distance, or across time. It investigates the impact of new communication technologies on domestic life and the changing nature of connection across a variety of family relationships, including couples, parents and children, adult siblings, and grandparents and grandchildren.

Family

The idea of "family" can no more be defined by a network of blood relations than the concept of "home" can be described as a physical building. At some level, we may think of family as a collective of partners, parents, children, grandparents, and

C. Neustaedter (✉)
School of Interactive Arts and Technology,
Simon Fraser University,
102 Avenue, 250-13450, V3T 0A3 Surrey, BC, Canada
e-mail: carman_neustaedter@sfu.ca

S. Harrison
Department of Computer Science and School of Visual Arts, Center for Human-Computer Interaction, Virginia Polytechnic Institute and State University,
2202 Kraft Drive, Blacksburg, VA 24060, USA
e-mail: srh@vt.edu

A. Sellen
Socio-Digital Systems,
Microsoft Research Cambridge,
JJ Thomson Avenue 7, Cambridge CB3 0FB, UK
e-mail: asellen@microsoft.com

C. Neustaedter et al. (eds.), *Connecting Families,*
DOI 10.1007/978-1-4471-4192-1_1, © Springer-Verlag London 2013

1

various other relations. But to stop here would be to gloss over what we really mean when we talk about being part of a family, spending time with family, or making a family home. These richer, everyday concepts point to a much more nuanced and profound idea of what a family is. When seen in these terms, it is clear that the notion of family is to some extent an aspiration—something we strive to achieve and a goal that we aim toward. Furthermore, moving toward this goal requires effort—and sometimes a great deal of effort—to maintain family, to nurture it, and to adapt domestic life to its changing needs and unfolding circumstances. In short, family is something that we *do*, not something that simply *is*. More than this, the doing of family is never complete. It is always a "work in progress".

To say that families require work is perhaps no surprise to the average hassled, over-tired parent. On the other hand, it may appear somewhat grandiose to speak of family life as aspirational. Indeed, the work that constitutes family life is at once both mundane and of fundamental importance. Research in anthropology, sociology, and, more recently, Human-Computer Interaction (HCI) has shown how the "doing" of family can be seen in many of the ordinary things we carry out every day. Examples include housework (Martin 1984), shopping (Miller 1998); cooking (Grimes and Harper 2008), the arranging of objects in the home (Miller 2008; Kirk and Sellen 2010), and even in how we deal with family clutter (Swan et al. 2008). But this research also shows that through these seemingly unremarkable activities, something much more valuable is achieved. When it comes to domestic life, we are not just tidying things up, bringing in provisions, preparing food and so on. Rather, we are fulfilling our duty, showing affection and concern for those we care about, and making a home in which family identity is expressed and reinforced.

In the midst of this, and in fact underpinning all of these activities, is *family connection*: how families not just communicate with each other, but how they share their lives and routines, how they engage in social touch, and how they negotiate being together, or being apart. This is the central theme of this collection of essays. In it we look at families from the most intimate relationships between couples to the dynamics of the immediate family, to extended and even fractionated families through divorce. We look at the sharing of ordinary life and special events, and the doing of everyday chores as well as play and laughter. And we examine how families strive to stay connected when they are separated by long distances, but also when they live together.

In these endeavors, technology has historically played a central supporting role, and, in turn, technological development has been spurred on by the needs of family connection. In today's world, technological change seems faster than ever, not just from the perspective of changes in speed, networking capacity, storage, and the proliferation of devices and services, but in terms of the choice that it offers up for new ways of connecting with family.

All of this raises important questions for the impact of new technologies on family life. Will new technologies help strengthen the bonds that already exist, or will they complicate or accentuate tensions? Will it allow us to connect more widely with others we care about, or will a pre-occupation with far-flung connections simply mean less time for those who are here and now, and closest to us? The answers are not simple, and the impact of technology will not be neutral. What we can be sure of is a fascinating story that will unfold as new technologies evolve hand in hand with changes in domestic practice.

The Importance of Connection

So what then does it mean to be connected? Within the immediate family, it may mean the ability for families to communicate with each other to coordinate, share their experiences, mediate their relationship, maintain varying degrees of intimacy, and, simply put, feel closer to one another. Within the extended family, it may mean staying abreast of major life changes, health issues, or general locations such as when a family member might be in town to visit. Maintaining a connection may be easier for some family members than others and will vary greatly depending on how much value individuals place on staying connected, how "important" individuals are within one's social circle, and so on. There are also those who we are keen to stay aware of, those who are harder to stay in touch with because of busy lives and schedules, and, yes, by human nature, even those family members with whom we have little desire to stay connected.

The need to be "connected" is also highly dynamic. Immediate family members such as parents and children may have a constant need to be connected because their day-to-day activities and functioning depends on it. Yet extended family members may stray out of touch in the absence of new or exciting events that warrant communication. On the other hand, when major life events occur like weddings, graduations, and, sadly, even funerals, the meaning of connection changes and its importance elevates. This is not to say that connection is always good, however. There are times when being apart is as important as being together. For example, adult children who have moved away from home to go to college or live on their own may desire less connection than their parents try to achieve. Or, similarly, so-called "helicopter parents" may try to "overconnect" with their children as they try to remove obstacles from their children's paths.

Often of pivotal importance for staying connected is where family members are located. Past research suggests that most people prefer to connect with their close family members in person (Ling 2000; Hindus et al. 2001; Neustaedter et al. 2006; Greenberg et al. 2009; Tee et al. 2009). Yet not everyone is able to see their family members in person when they need to or want to. For family members who live together, connecting may occur within the same home or outside of the home between the various locations that people visit throughout the day, such as work or school. Family members who live apart must connect across distance where the distance might be small, such as across a city, or large, such as across the country or even the world. Because of this, family members have used varying "technologies" to connect with each other over distance. Prior to the dawn of the Internet, if opportunities for face-to-face communication or exchange were limited or not possible, families primarily connected with each other through the telephone and postal mail system.

For example, while at work, parents called each other using the phone to coordinate children's activities for the evening. Children similarly called their grandparents who live across country to tell them about their school or extracurricular activities. Families also relied on the postal system to send letters, cards, and other greetings to their remote family members where they would feel more connected despite the medium's less-timely communicative nature.

When we consider the notion of family connection, it is also clear that the value of connecting for families is quite different from that of connection in the workplace. In the fields of HCI, and more especially Computer-Supported Cooperative Work (CSCW), research in the workplace context and media space literature has shown how connection is rooted in moving information, coordinating tasks and negotiation (Bly et al. 1993; Harrison 2009). Mediated connection in the workplace is motivated by needs of workflow, projects, and organizational structures. As this literature developed, it became apparent that sociality was an important component of mediated workplaces, but "feeling connected" was not a priori a driver of system design. In contrast, the domestic realm focuses on connection for its own sake. The state of being "together" as a member of a household, as a member of a family, as a member of a couple bears only a superficial similarity to being together on a project with co-workers. "Connection" is part of the identity of being a family. The informational content of "connection" is often secondary to the reassurance of awareness and presence.

This raises the issue of how this new orientation to home life will build on existing research and literature. There are many reasons why work-related research will remain relevant. One is that the boundary between work and family is increasingly blurred, which means that some practices from the workplace will increasingly find their place in the home. Another is that the technologies and practices around connection that we focus on in this volume may be based on workplace technologies that have become re-worked in the family context. And finally, the contrast between family and work expands the definition of "connection" for both. There may be parallels with personal computing which began as a home-based phenomenon (although resting on technology from the workplace); personal computing spread into the workplace and reconfigured work, the workplace and working relations. So, will mediated connection in the family realm become a distinctively separate kind of technology from that in the workplace? If it does, how will it reconfigure work? It is important to both track the phenomenon and actively explore alternatives.

A Changing World

In order to track the changes going forward, it is important to have line of sight into the past. When we do, and as others have discussed (Harper 2010), we find that new technologies tend not to replace the old ones, but instead they add to the palette of possibilities. In turn, old technologies find their place, and sometimes evolve in response to these new niches. As a case in point, in present day, the telephone and postal mail system are still used by family members to connect with each other, but the meaning of a phone call, or a letter or card has changed. In fact a paper card may in some ways be more special by the very fact that it is so much easier to send something digitally. Email, too, may now be seen as quite old fashioned in some ways, and may even be eschewed by younger generations who insist that social networking tools are the way to connect with friends. Yet email remains fundamental to how we do our work, and even teenagers recognize that email may be useful for communicating with teachers or doing other "work-related" things. In a sense, telephone, paper mail and email are all continuing to find their place in the world, even though that place is constantly evolving.

Driving these changes is a host of new technologies that provide additional means for connection, many of which have been brought about by the need to stay connected to family. Mobile phones, for example, have dramatically changed the nature of family connections by making family members accessible in nearly any place, at any time (Ling 2000). The Internet and mobile wireless networks have caused mail systems to evolve to support the exchange of messages almost instantly via email, instant messaging, and text messaging between friends and family. Research in computer-mediated communication has explored the ways in which family members use these communication, awareness, and interaction technologies, as well as how to best design family communication technologies of the future that can seamlessly bring people together and help them feel connected regardless of their location.

For example, we see focal points on bringing together grandparents and grandchildren in the moment through video communication systems (Judge et al. 2010; Kirk et al. 2010; Ames et al. 2010; Judge and Neustaedter 2010). This occurs in the context of the home (Sellen et al. 2006; Kirk et al. 2010; Ames et al. 2010; Judge and Neustaedter 2010) and also while family members are mobile (O'Hara et al. 2009). Research has also looked at how parents who long to stay aware of their adult children as they grow up and leave 'the nest' stay connected with them (Tollmar and Persson 2002; Plaisant et al. 2006; Lindley et al. 2009). The reverse has also been studied where researchers have investigated how adult children stay connected with their aging parents, often to ensure their health and welfare is fine (Mynatt et al. 2001). Together, this research and more has resulted in a number of technological advances for bringing together and connecting family members, including messaging systems, information appliances, and mobile applications (e.g., Strong and Gaver 1996; Hindus et al. 2001; Hutchinson et al. 2003; Romero et al. 2006; Sellen et al. 2009).

As is evident, technology is playing an increasing role in mediating family relationships. Here the social, cultural, and technological issues are increasingly rich and complex, as family members must understand what technologies are available to them, learn how to use them, and adapt them into their existing communication routines and practices. This brings challenges with technology usability where family members, such as children (Ames et al. 2010) or older adults (Mynatt et al. 2001; Lindley et al. 2008), might struggle with understanding how to get a technology to "do what they want." Family members face issues with time zone separation where they must figure out how to best "schedule" or "time" their communication with those afar (Ciao et al. 2010). Family members must also balance their needs to stay connected with privacy issues of revealing or sharing too much information, or being "too connected" (Judge et al. 2010; Birnholtz et al. 2010). We also see issues with social isolation where individuals may want more connection with their family members, yet they are unable to achieve such connections for a variety of reasons (Grenade and Boldy 2008). This list could certainly go on and on, which is why the study of "connecting families" is of increasing importance in present day.

Beyond this, we are now seeing an increasing trend, which further brings this research space to the forefront. Computer-mediated communication technologies for families are now moving out of the research lab and into actual everyday practice. In fact, one might argue that some technologies, such as video-communication systems, are finding stronger purchase and presence in the home environment than in the workplace. These computing technologies are rapidly changing the way families can communicate, coordinate, and connect with others through readily available (and often free) applications, such as Google Talk, Skype, or Apple's FaceTime. The accessibility and proliferation of these applications means that family members are increasingly faced with new mechanisms to reach out and connect with their family and friends. For this reason, technology is now rapidly reconfiguring the way we think about and design for domestic spaces and domestic life. As it does so, researchers now must directly confront issues of family relations and the subtle negotiations that are part of that realm.

Purpose of the Book

In what follows, we bring together a collection of chapters that constitutes both a diverse overview of research into technologies for connecting families, and one that offers a comparative guide both in terms of the relationships under scrutiny, and the technologies that are evaluated. Specifically, it brings together studies with various relationship dynamics ranging from intimate partners to extended family such grandparents and grandchildren. It also explores a variety of technological solutions, including mobile devices, information appliances, and computer applications; media such as text, video, and audio; and, function where it explores awareness, interaction, and other forms of communication. The goal is to bring these case examples together in order to allow readers to draw their own perspectives and conclusions that cross relationships and technology boundaries.

The book can be used in a variety of ways. First, it can act as a tool for courses focused on studying domestic relationships, routines, or technology usage. In this way, the entire book, or specific chapters can be used as studies of particular facets of "connecting families". Second, it can serve as a resource for those conducting research in the area of family communication that brings together both state of the art and foundational literature, including the chapters themselves as well as the works referenced within them. This should aid those who are studying varying family relationships including connecting partners, parents and children, children, and grandparents and grandchildren. It can also aid those studying various technology or communicative or media forms, such as video-based communication or messaging systems. Third, most of the chapters have important implications for new technologies we might design, both in terms of underlying concepts and the requirements for those technologies. As we shall see, the needs of different "user groups" as defined by their relationships (whether we are talking about couples, children with peers, intra-family relationships and so on) may be quite different. This in turn gives guidance as to what these different groups might value, and how technology might best support those values. The book then can be used by those who may have a more applied rather than theoretical focus.

Overview of the Book

The book is partitioned into three main sections based on the varying relationships that shape the nature of communication and the technologies that underpin it: couples and partners, immediate families and children, and the extended, distributed family.

1. Couples We start with what is often the core of a "family," the couple, to look at and understand the impact of technology design on couple relationships, communication, and feelings of closeness. In couple relationships, connection is often of the utmost importance to keep partners together, maintain the intimacy of the relationship, and coordinate day-to-day activities.

This section begins with Branham and Harrison's chapter on designing for Collocated couples that puts forward the notion of "couple-centered design". Here the emphasis is on designing technologies with "the couple" as core user as opposed to many designs, which focus solely on ensuring usability and usefulness for the individual. In this chapter, Branham and Harrison present a variety of technologies that have been designed over the years for both collocated and distributed couples along with their prototype of a Diary Built for Two and discussions of how it can promote deep interpersonal sharing for collocated couples. Together, this presents a framework for how one can think about couple-centered design.

Building on this, we then narrow the focus and move to Greenberg and Neustaedter's chapter on intimacy in long-distance couple relationships. This chapter explores the unique way in which long-distance partners have appropriated "off-the-shelf" video chat systems like Skype to stay connected. In many cases, these couples are using video chat systems akin to media spaces from the workplace (Harrison 2009) where the video and audio links are left on for extended periods of time. Here couples value being able to connect their distributed residences with the technology to create a shared sense of place. This "shared living" across distance helps them share life, experiences, and intimacy, despite social and technical challenges created by the technology.

2. Immediate Families & Children Section Two moves on to studies of immediate families that have expanded beyond just the individual or couple to include children. Here we present chapters that investigate the design of technology to connect families as a part of their everyday living practices within the home, or connecting across homes. This includes parents connecting with their children as well as the situations that arise when children want to connect with their friends over distance. The emphasis is on connection for communication, coordination, and play.

We begin the section with Schatorje and Markopoulos's chapter on intra-family messaging that explores "connecting" in households comprised of parents and teenage children. The chapter's emphasis is on designing family technologies in a flexible manner such that existing routines can easily migrate to new communication technologies. To this end, they describe the design evolution of Family Circles, a messaging system that allows family members to leave audio recordings for each other on tokens that can be placed throughout the home. This migrates family practices of leaving handwritten messages for one another to a new technological form.

Next, we examine parent-child relationships where communication and interaction has been complicated because of divorce. In these cases, the "simple" situation presented in the preceding chapter where parents and children all reside in the same household is no more, and at least one parent lives apart from his or her child. To this end, Yarosh and Abowd's chapter on enriching virtual visitation describes the challenges that divorced families face when trying to connect parents and children across households and the opportunities for designing technologies to support them. They present the design of the ShareTable that allows parents to interact and engage with their children over distance with the aid of an audio and video connection. The chapter also emphasizes the many pragmatic and challenging issues that can arise when moving a prototype out of the research lab and into the home for real usage.

Following this, we look more specifically at connecting children to investigate how technology can be used to mediate child-to-child relationships, such as friends or cousins, over distance. This is one part of domestic life that parents must often account for and facilitate in order to ensure their children have their social skills enriched and nurtured. As relationships in society become increasingly mediated by technology, so too do those amongst children who often desire to connect with their friends over distance. Inkpen's chapter explores how both asynchronous and synchronous video chat systems can support children playing and interacting over distance and the advantages that each brings forth. This includes the presentation of three prototype systems, Video Playdate, IllumiShare, and VideoPal.

3. The Extended, Distributed Family Lastly, we move beyond the immediate family to explore connections between extended, distributed family members. This includes connections between adult children and their parents, grandparents and grandchildren, and adult siblings. Here family members have grown older, moved away from "home," and forged "new" families. Yet they still have needs to connect with their existing family members. In these situations, we often see the most diversity in terms of connecting. The needs for connecting may be highly dynamic and change depending on life events. They may also be much more discretionary if relationships are not particularly strong, or there could be a real desire by family members to connect more because they miss their extended family.

First, Cao's chapter on connecting families across time zones sets the framework for thinking about extended family connections. He describes the many challenges that parents and adult children, as well as siblings, face when trying to connect across distance when time zones come in to play. In these situations, family members must often coordinate, plan, and schedule interactions when each person may have a very different notion of time, day, night, etc. Cao juxtaposes the importance of synchronous and asynchronous communication in these situations.

Next we focus in on one type of technology that can connect extended family members who are distributed: video conferencing in the form of a domestic media space. Judge, Neustaedter, and Harrison's chapter explores the design and usage of two such systems, the Family Window and Family Portals, and how parents and adult children, grandparents and grandchildren, and adult siblings used the messaging features within these systems to stay connected. Some family members were separated by time zones, and all were separated by distance. Here the notion of connection refers to the ability for the systems to make family members feel close to one another and aware of their day-to-day activities.

Following this, we narrow in on the grandparent-grandchild relationship more deeply for the final two chapters in this section. First, Ballagas, Kaye, and Raffle's chapter explores "connected reading" and how video communication systems focused on play and reading can support grandparent interactions with young grandchildren. They present three systems, Family Story Play, Story Visit, and People In Books, where each embeds video within a storybook in a unique way. The act of tying family connection to an activity that children love, namely reading, allows grandparents to share longer, more meaningful time with their grandchildren than other more traditional technologies (e.g., the phone).

Lastly, Moffatt, David, and Baecker's chapter takes a step back from the previous chapter to explore grandparent and grandchild relationships more holistically to understand their role throughout life as they grow and evolve. This includes relationships between grandparents and young grandchildren as previously discussed, as well as teenage, and even adult grandchildren. They illustrate how a variety of technologies can support such relationships, including those focused on shared reading with young children, shared stories about family history for older grandchildren, collaborative reading for situations where grandparents have difficulties reading, and biographies to act as a catalyst for conversation.

Book Themes

Beyond the explicit structure that we have presented above, there are several themes that resonate throughout chapters within the book and spread across multiple sections. At a surface level, this includes designing for varying age groups and family roles. Yet, beyond this, the chapters provide an additional understanding of the ways in which family connection has been studied. Some of the more prominent themes include:

Methodologies The book presents chapters that include a range of methodologies for studying family connection. This includes interviews to understand existing family practices and guide new designs (Chaps. 3–5, 7, and 9); information probes to inform and inspire design (Chap. 4); iterative design and prototyping of new technologies (Chaps. 2, 4, 5, 8, 9, and 10); field trials of prototype technologies coupled with interviews to more deeply understand usage (Chaps. 2, 4, and 8); and laboratory studies aimed at guiding design and understanding new technology usage (Chaps. 5 and 6). As can be seen, most often, family connection is studied using exploratory, qualitative methods where there is an emphasis on studies performed in homes or the field. However, there also exist studies that are more quantitatively focused or occur in a controlled, lab setting. The challenge is being able to create a natural and realistic setting that replicates domestic spaces or practices.

Design-Research Lifecycle Related to methodologies, we also see chapters that focus on varying points in the design-research lifecycle. Some are focused on early design research that explores a particular type of relationship, technology area, or family practice in the form of gathering design requirements or providing descriptive accounts of domestic life (Chaps. 3 and 7). This knowledge can then be used as a basis for designing future technologies. Some chapters are focused on the actual design and evaluation of technologies where a prototype system is created, often through iterative design, and then evaluated either in the field or lab (Chaps. 6, 8, and 10). Other chapters describe larger portions of the lifecycle and include stages of requirements gathering, design, and evaluation (Chaps. 4, 5, and 9).

Technological Medium A strong focus across chapters is also the technological medium being explored. Family connection can be supported in many ways through technologies and researchers have explored a variety of options. One of the most predominant mediums, at least explored in this book, is the use of video connections that are sometimes coupled with audio links (Chaps. 3 and 5 through 10). In addition to this, we also explore audio messaging (Chap. 4) and textual-based communication (e.g., diaries, handwritten messages, stories) (Chaps. 2 and 7 through 10). Across these mediums, some chapters are focused on synchronous communication where family members can connect in real time (Chaps. 3 and 5 through 10), while others explore asynchronous communication spread over time (Chaps. 2, 4, 6–8, and 10).

We hope that readers will latch on to these themes and others as they explore the research space presented in this book. This may be especially valuable for those using the book as part of a design or human-computer interaction course, or for researchers learning more about the topic of "family connection".

References

Ames, M., Go, J., Kaye, J., Spasojevic, M. (2010). Making love in the network closet: the benefits and work of family videochat. *Proceedings of the CSCW 2010* (pp. 145–154). New York: ACM.

Birnholtz, J., Jones-Rounds, M. (2010). Independence and interaction: understanding seniors' privacy and awareness needs for aging in place. *Proceedings of the CHI 2010*. Atlanta: ACM.

Bly, S., Harrison, S., Irwin, S. (1993). Media spaces: bringing people together in a video, audio, and computing environment. *Communications of the Association of Computing Machinery, 36*(1), 28–45.

Ciao, X., Sellen, A., Brush, A. J., Kirk, D., Edge, D., Ding, X. (2010). Understanding family communication across time zones. *Proceedings of the CSCW 2010* (pp. 155–158). New York: ACM.

Greenberg, S., Neustaedter, C., Elliot, K. (2009). Awareness in the home: the nuances of relationships, domestic coordination and communication. In P. Markopoulos, B. Ruyter de, W. Mackay (Eds.), *Awareness systems: advances in theory, methodology and design*. New York: Springer.

Grenade, L., Boldy, D. (2008). Social isolation and loneliness among older people: issues and future challenges in community and residential settings. *Australian Health Review, 32*(3), 468–479.

Grimes, A., Harper, R. (2008). Celebratory technology: new directions in food research for HCI. *Proceedings of the CHI'08*. Florence: ACM.

Harper, R. (2010). *Texture: human expression in the age of communications overload*. London: MIT.

Harrison, S. (2009). *Media space: 20 + years of mediated life*. New York: Springer.

Hindus, D., Mainwaring, S. D., Leduc, N., Hagström, A. E., Bayley, O. (2001). Casablanca: designing social communication devices for the home. *Proceedings of the Conference on Computer-Human Interaction (CHI 2001)* (pp. 325–332). New York: ACM.

Hutchinson, H., Mackay, W., Westerlund, B., Bederson, B., Druin, A., Plaisant, C., Beaudouin-Lafon, M., Conversy, S., Evans, H., Hansen, H., Rouseel, N., Eiderback, B., Lindquist, S., Sundblad, Y. (2003). Technology probes: inspiring design for and with families. *Proceedings of the Conference on Computer-Human Interaction (CHI 2003)* (pp. 17–25). CHI Letters 5(1). New York: ACM.

Judge, T. K., Neustaedter, C. (2010). Sharing conversation and sharing life: video conferencing in the home. *Proceedings of the CHI 2010*. New York: ACM.

Judge, T. K., Neustaedter, C., Kurtz, A. F. (2010). The family window: the design and evaluation of a domestic media space. *Proceedings of the CHI 2010*. New York: ACM.

Kirk, D., Sellen, A. (2010). On human remains: value and practice in the home archiving of cherished objects. *ACM Transactions on Computer-Human Interaction, 17*(3).

Kirk, D., Sellen, A., Cao, X. (2010). Home video communication. *Proceedings of the CSCW 2010* (pp. 135–144). New York: ACM.

Lindley, S. E., Harper, R., Sellen, A. (2008). Designing for elders: exploring the complexity of relationships in later life. *BCS-HCI '08 Proceedings of the 22nd British HCI Group Annual Conference on People and Computers: Culture, Creativity, Interaction* (Vol. 1, pp. 77–86).

Lindley, S. E., Harper, R., Sellen, A. (2009). Desiring to be in touch in a changing communications landscape: attitudes of older adults. *Proceedings of the 2009 SIGCHI conference on Human Factors in computing systems (CHI 2009)* (pp. 1693–1702). New York: ACM.

Ling, R. (2000). Direct and mediated interaction in the maintenance of social relationships. In A. Sloane F. Rijn van (Eds.), *Home informatics and telematics: information, technology and society* (pp. 61–86). Boston: Kluwer.

Martin, B. (1984). Mother wouldn't like it: housework as magic. *Theory, Culture & Society, 2*(2), 19–36.

Miller, D. (1998). *A theory of shopping*. Cambridge: Polity Press.

Miller, D. (2008). *The comfort of things*. Cambridge: Polity Press.

Mynatt, E., Rowan, J., Jacobs, A., Craighill, S. (2001). Digital family portraits: supporting peace of mind for extended family members. *Proceedings of the Conference on Computer-Human Interaction (CHI 2001)* (pp. 333–340). CHI Letters 3(1). New York: ACM.

Neustaedter, C., Elliot, K., Greenberg, S. (2006). Interpersonal awareness in the domestic realm. *Proceedings of the OzCHI*. New York: ACM.

O'Hara, K., Black, A., Lipson, M. (2009). Media spaces and mobile video telephony. In S. Harrison (Ed.), *Media space: 20 + years of mediated life* (pp. 303–323). New York: Springer.

Plaisant, C., Clamage, A., Hutchinson, H., Bederson, B., Druin, A. (2006). Shared family calendars: promoting symmetry and accessibility. *Transactions on Computer Human Interaction, 13*(3), 313–346.

Romero, N., Markopoulos, P., Baren, J., van., Ruyter, B., de., Jsselsteijn, W., Farshchian, B. (2006). *Connecting the family with awareness systems, personal and ubiquitous computing* (Vol. 11, pp. 299–312). New York: Springer.

Sellen, A., Harper, R., Eardley, R., Izadi, S., Regan, T., Taylor, A., Wood, K. (2006). Situated messaging in the home. *Proceedings of the CSCW 2006*. New York: ACM.

Sellen, A., Taylor, A. S., Kaye, J., Brown, B., Izadi, S. (2009). Supporting family awareness with the whereabouts Clock. In P. Markopoulos, B. Ruyter de W. Mackay (Eds.), *Awareness systems: advances in theory, method and design*. New York: Springer.

Strong, R., Gaver, B. (1996). Feather, scent, and shaker: supporting simple intimacy. *Proceedings of the CSCW'96* (pp. 16–20). New York: ACM.

Swan, L., Taylor, A. S., Harper, R. (2008). Making place for clutter and other ideas of home. *ACM Transactions on Computer-Human Interaction, TOCHI, 15*(2).

Tee, K., Brush, A. J., Inkpen, K. (2009). Exploring communication and sharing between extended families. *International Journal of Human-Computer Studies, 67*(2), 128–138.

Tollmar, K., Persson, J. (2002). Understanding remote presence. *Proceedings of the NordiCHI 2002* (pp. 41–49). Arhus: ACM.

Part I
Couples

Chapter 2
Designing for Collocated Couples

Stacy Branham and Steve Harrison

Abstract Though the design of technologies for couples has been thriving for well over a decade now, the products made for and the needs of couples examined in HCI research are surprisingly narrow. Overwhelmingly they are for *partners at a distance* and lightweight interactions that can best be described as *abstracted presence*. Towards moving couples technologies into broader waters and guiding exploration of the many other facets of couplehood, we propose an expanded couples design space that includes technologies for *local partners* and *deep interpersonal sharing*—hitherto underexplored design concerns. We then show that the creation of these new spaces can be motivated by the needs of couples as characterized by couples experts and present an example of a new technology that embodies these. Finally, we draw from experience with couples in the field to identify research and design considerations regarding gender, power, values, and ethics.

Introduction

When we tell someone for the first time that we design technologies for couples, they often ask the question "why design for couples; what's so special about *them* as opposed to just anyone?" This question is one that strikes to the core of the burgeoning research on domestic technologies, though it is one that has yet to be adequately addressed in HCI. When we design for the home, are we designing for the individual, the couple, the children, or the family? Mightn't close friends also

S. Branham (✉) · S. Harrison
Department of Computer Science and School of Visual Arts, Center for Human-Computer Interaction, Virginia Polytechnic Institute and State University,
2202 Kraft Drive, Blacksburg, VA 24060, USA
e-mail: sbranham@vt.edu

S. Harrison
e-mail: srh@vt.edu

C. Neustaedter et al. (eds.), *Connecting Families,*
DOI 10.1007/978-1-4471-4192-1_2, © Springer-Verlag London 2013

benefit from technologies designed for domestic relationships? And, don't couples sometimes want to be treated as individuals or friends or fill any number of other roles at various times? In essence, *what makes a couple a couple* (or a family a family, and so on)? These questions are central to the spirit of this book and this chapter in particular.

Where Couplehood Meets Technology: A Personal Example

Stacy Branham: I was at first unaware of the need for what I now call couple-centered-design—that is, until the need hit close to home. Below, I share an example from my experience with a technology that seems to have neglected my needs as a partner in a relationship. The culprit is Netflix circa late 2010 (netflix.com). When my partner of 3 years (now my husband) and I decided to subscribe to this popular online movie rental service, it was clear to us that it was something we wanted jointly—a technology for our couplehood—so we billed it to our joint bank account. But, as we would soon discover, Netflix is (perhaps by profit-seeking design) clearly not cut out for two.

The primary source of my frustration with Netflix is that there is no good way for my husband and me to find movies that we both like, or even that we individually like. Netflix's movie recommendation engine requires us to rate movies we've already watched, but the system does not allow us to enter different ratings for the same movie, nor does it allow us to distinguish between my ratings and his. This means that movie recommendations reflect the tastes of the partner who was first to rate each movie or who has made the most recommendations (my partner, in both cases). Consequently, Netflix has brought with it a host of minor yet new arguments over such pressing matters as "why is Pride and Prejudice rated 5 stars?" and "why is our recommendation list littered with kung fu titles?"

It appears that Netflix is not completely unaware that the service will be used by families. Netflix has a "profiles" feature through which I was able to create my own profile, rate my own movies, and receive my own recommendations. But, after some awkward interactions, I began to see this as a feature intended for children rather than an account co-owner. If I want to add a movie to the queue of DVDs that will be sent to our house, Netflix suggests that I "contact [my] account manager" to ask him to add the movie. I am also not allowed to add a movie to the queue of movies that can be streamed instantly over the Internet. And, if I want to switch back to Jason's profile to gain access to these features, I am prompted to log in again. Switching to my profile from his, on the other hand, is a simple a matter of clicking a button. So, access to personal information and key account features, including actually watching a movie, is not reciprocal.

Part of my problem with Netflix is that I cannot easily find and access movies that fit my personal preferences in the same way that my partner can. What's more, Netflix may be missing an opportunity to enhance my sense of connection with my partner via foregrounding our shared movie tastes and helping us find movies that we can both enjoy. But there is another, more subtle issue: the way we currently use the system casts me as a subordinate user. Jason's movie preferences take precedence, and I am not granted reciprocal access to our stored data and system controls. As an equal payer and partner, this is not acceptable and has often resulted in minor tension, although tension nonetheless, in our relationship[1].

---

[1] Note that couple-centered design is not confined to delivering constructive experiences and avoiding destructive ones, as these terms may be variously defined. It is instead about understanding how technology can interact with couplehood and designing accordingly.

I do not wish to suggest by sharing this example that all couples would respond to Netflix in the way my partner and I have. I simply wish to show that technologies like Netflix interact with couplehood, an observation that others have also made (Wilson 2009). Consequently, design for the individual and design for the couple are not always the same and should likely result in different interactive configurations. Furthermore, design for the family, which Netflix seems to have attempted with its profile feature, must also consider design for the couple. There is indeed something about couples that makes them different, but exactly what that something is has yet to be thoroughly explored in HCI.

In This Chapter

In this chapter, we present an incremental contribution that addresses the question raised in our opening comments—"what makes a couple a couple?" We do so by adding to the current understanding of what's "in" in terms of couplehood design concerns. To start, we propose an expanded couples design space that includes technologies for *local partners* and *deep interpersonal sharing*. We then show that these two design locales can be motivated by the perspectives of couples experts, and we present an example of a new technology that fits within the expanded design space. Finally, we use examples from my experience with couples to motivate a new set of couplehood design considerations regarding gender, power, values, and ethics.

The Existing Couples Design Space

Technology for couples is by no means an untapped design space. Designs began to emerge within HCI as early as 1996 (Strong and Gaver 1996), and through an ongoing literature review, over 40 system concepts or implementations directed specifically toward couples have been identified (see Table 2.1). These systems can transmit digital kisses, touches, hugs, hand-holds, and kicks. They can send signals that smell, float, light up, warm up, vibrate, spin, play music, and more. Yet, even amidst this diversity, a broad perspective reveals that these technologies fall within a relatively narrow band of a much larger potential couples design space. In the following paragraphs, we briefly summarize some of the more notable designs towards mapping out new frontiers of the design space for couples.

Notable Couple Technologies

Feather and Scent (Strong and Gaver 1996) were two of the first couple technologies to be published in HCI. Both are targeted at relationships in which one partner is traveling while the other remains at home. Feather is composed of two physical

Table 2.1 Chronological list of 40 couple technologies characterized by design motivation

List of Couple Technology Designs

technology	for distant or local partners?	for abstracted presence or deep interp. sharing?
Feather [46]	distant	abstracted presence
Scent [46]	distant	abstracted presence
The Bed [24]	distant	abstracted presence
inTouch [15]	distant	abstracted presence
LumiTouch [16]	distant	abstracted presence
The Sensing Beds [25]	distant	abstracted presence
Habitat [41]	distant	abstracted presence
Honey I'm Home [31]	distant	abstracted presence
How Do I Love Thee? [33]	distant	abstracted presence
Hand-holding [33]	distant	abstracted presence
Love Egg [33]	distant	unspecified
Telesquishy [7]	distant	abstracted presence
Kiss Communicator [8]	distant	abstracted presence
Hug Over a Distance [39]	distant	abstracted presence
SecretTouch [49]	distant	abstracted presence
i.Fuzz [49]	local	unspecified
SynchroMate [49]	unspecified	abstracted presence
Anemo [40]	distant	abstracted presence
Air [40]	distant	abstracted presence
Tok Tok [9]	distant	abstracted presence
Tug Tug [9]	distant	abstracted presence
I Just Clicked... [32]	distant	abstracted presence
Lover's Cups [18]	distant	abstracted presence
Hug Shirt [5]	distant	abstracted presence
Digital Selves [26]	distant	abstracted presence
ComSlipper [17]	distant	abstracted presence
SyncLamp [48]	distant	abstracted presence
SyncTrash [48]	distant	abstracted presence
SyncSky [48]	distant	abstracted presence
Duofone [4]	local	unspecified
Hello There [34]	distant	deep interpersonal sharing
Traveling Book [34]	distant	deep interpersonal sharing
Our Day [34]	distant	deep interpersonal sharing
Mutsugoto [28]	distant	abstracted presence
Daily Temp. Reading [11]	unspecified	deep interpersonal sharing
MissU [37]	distant	abstracted presence
Digital Kick [20]	local	unspecified
Fix a Fight [12]	local	deep interpersonal sharing
CoupleVIBE [14]	distant	abstracted presence
Aura [13]	unspecified	abstracted presence

artifacts, one that resides as a stable piece of furniture in the home—a glass vase containing a feather—and the other that stays with the traveling partner—a picture frame. When the remote partner strokes the frame, the feather in the vase is briefly floated in the air by a small fan at the base of the fixture. Similarly, Scent is comprised of two objects, one being a picture frame. In place of the vase, Scent introduces an aluminum bowl with a heating element. When the remote partner handles the picture frame, the heating element vaporizes essential oils contained in the bowl, filling the air with a lingering fragrance. These two systems foreground the subtlety of intimate communication and the value of implicit, non-verbal, symbolic interaction.

inTouch (Brave and Dahley 1997) is one of the first couple-targeted systems to seek simulation of touch towards more intimate communication, though its designers were expressly against simple mimicry of existing physical forms of human-to-human touch. Instead, they designed a pair of devices outfitted with three rollers each. Two distant partners might communicate using these devices by rolling their hands over their respective device. When one of the rollers is rotated, the corresponding roller on the remote device also rotates. Like Feather and Scent, inTouch is characterized by "subtle and abstract…interaction" and a "lack of ability to pass concrete information" to one's partner.

LumiTouch (Chang et al. 2001) is another system designed for geographically-separated partners. LumiTouch consists of two picture frames. When one partner handles their picture frame, the remote partner's frame illuminates with colors that correspond to where, how hard, and how long the frame is squeezed. The authors suggest that the abstract communication supported by the system may be able to take on more nuanced meanings via creation of an "interpersonal language;" "the combination of colors and force allow[s] a grammar, while the duration of squeeze provide[s] syntax for creative interpersonal dialect between two people."

Hug Over a Distance (Mueller et al. 2005) is a system that supports tactile interactions that simulate hugs between partners that cannot be physically copresent. Its designers were inspired to create haptic experiences for couples that act as "emotional pings"—interactions akin to "small 'I love you' text messages." The intention is for each partner to wear a vest that can fill with air to mimic the sensation of a hug. The "hug" can be initiated by either partner by making a hug gesture; the other's vest will then fill with air until the hug is released by the initiator.

I Just Clicked to Say I Love You (Kaye 2006)—like Feather, Scent, and Hug Over a Distance—offers one-bit communication for couples in long-distance relationships. The system runs on each partner's personal computer. When one partner clicks on the circle displayed on their screen, the corresponding circle on the other's screen turns red, fading in color over time. Each partner is able to view the color of their significant other's circle. As proposed by Chang et al. (2001) regarding LumiTouch, field trials with I Just Clicked to Say I Love You suggest that even one-bit communication can generate rich interpretations; "a single bit of communication can leverage an enormous amount of social, cultural and emotional capital, giving it a significance far greater than its bandwidth would seem to suggest."

Table 2.1 lists these technologies and 34 others that have been identified through a review of the HCI and related literature, as well as through a web search of design sites. The inclusion criterion for this list was simple: did the designers explicitly identify couples as a target user group? Certainly, there are many systems developed for families, close friends, and other types of users—as those described in most chapters of this book—that couples might readily co-opt and find useful. We have intentionally limited the scope of this list to aid the task of considering what design motivations and design outcomes become apparent when designers take couples as their target users. In the coming paragraphs we will explore just that: what can current couple technology designs reveal about the prevailing design assumptions regarding who couples are and how technology can serve them?

The Imagined Couple

Looking at the current couples design space can provide insight into the prevailing assumptions about who couples are and what they need (or perhaps more interestingly, don't need) from their technologies. To this end, we have characterized the designs described above and those included in Table 2.1 according to two overarching design motivations as reported by the designers: *connecting partners at a distance* and supporting intimacy and connectedness via *abstracted presence*. While there are undoubtedly other ways to characterize these technologies, this particular characterization is the one that seemed most salient and pervasive. We will describe these two recurring design motivations in more depth below.

The first core tenant of the collective design thinking for couple technologies is the notion that couples separated by distance are most likely to benefit from technological mediation. Few papers have done much more to describe the situation of couples in this user group than identify that they are "separated by distance," "geographically separated," or "in long-distance relationships." For Chang et al. (2001), being distant means "living or working separately," and for Strong and Gaver (1996), distant partners may be temporarily separated for travel. In the study of CoupleVIBE (Bales et al. 2011), distant participants had "been apart for 6 months or more and were separated by at least 400 miles." Beyond small hints like these, no one has actually defined what it means to be partners at a distance. This may in fact suggest that the meaning of being distant is difficult to pinpoint, and furthermore that an ambiguous definition may be all that is needed; after all, the absence of a definition these past 15 years has certainly not prevented these designers from envisioning a number of couple-centered technologies. Table 2.1 shows that 33 of the 40 technologies can be categorized as being motivated by the distant partner problem[2]. And, a glance at the table of contents of this book echoes this trend; the problem of couples and families separated by distance is a highly compelling and frequently addressed one.

---

[2] Three technology designers did not specify their target user group in reference to distance.

The second core tenant of the collective design thinking for couple technologies is the notion that couples (often those separated by distance) prefer to use technology to communicate via "abstracted presence." Abstracted presence, as defined by Dodge (1997), is about providing "intimate, non-verbal inter-personal communication." A technology that supports abstracted presence is characterized by "its ability to become a shared virtual space… through aural, visual, and tactile manifestations of subtle emotional qualities" (Dodge 1997). Though abstracted presence is a term thus far used only to describe The Bed (Dodge 1997), most other couple-centered technologies fit this definition and are even described by their creators in similar terms. For example, when Strong and Gaver (1996) describe the subtlety of everyday sociality as the inspiration behind Feather and Scent, they explain that there is "no explicit communication, but instead a myriad of more basic visual, auditory, and tactile links are shared." Furthermore, they note that "the concern is not to exchange information, but rather to express mood and emotion." Others have also picked up on this trend. After their literature review of some of the technologies listed in Table 2.1, Davis et al. (2007) say the following: "what these technologies have in common is that they aim to evoke intimate reactions by relying on materials and abstract representation." Similarly, Lindley et al. (2009) note that "…technologies designed to mediate personal relationships are often lightweight. They afford a type of contact that is sufficiently vague to be interpreted as a show of tenderness, while precluding the communication of specifics." Table 2.1 shows that 31 of the 40 technologies can be categorized as being motivated by the desire to support abstracted presence[3].

Expanding the Design Space

Characterizing the couples design space as largely centered about two foci, distant partners and abstracted presence, begs the question "is that really all there is to couples?" Are there not other needs and other technologies to meet those needs in the couples space? This characterization can also lead us to some answers if we use it as a sort of scaffolding to envision new design opportunities. We can imagine, for example, that distant partners and abstracted presence are two ends of intersecting continua that transition into *local partners* and *deep interpersonal sharing*, respectively. By extending the distant partners design locale along a spectrum that leads to local partners and likewise extending the abstracted presence locale along a spectrum that leads to deep interpersonal sharing, we can visualize three underexplored design opportunities: abstracted presence for local partners, deep interpersonal sharing for local partners, and deep interpersonal sharing for distant partners. Notably, most couples technologies currently occupy the spaces defined by distant partners and abstracted presence, and few sit within the new spaces (Fig. 2.1).

---

[3] Four technology designs did not restrict communication to abstracted presence.

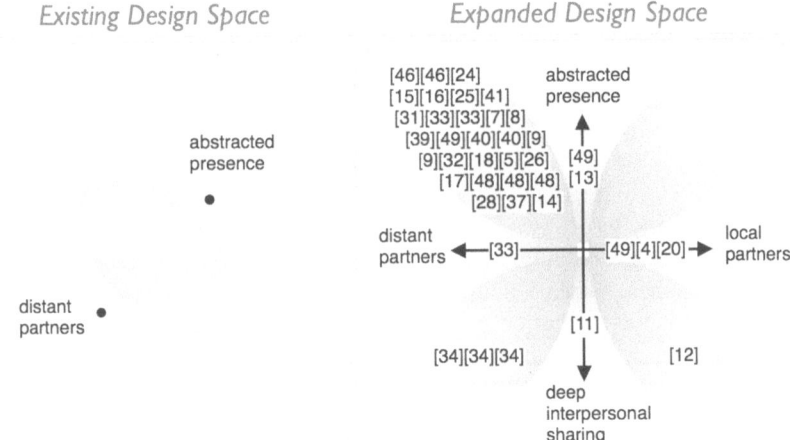

abstracted presence — *abstract; sensory (e.g., aural, tactile, visual); emotional communication; ephemeral, lightweight*
deep interpersonal sharing — *grounded; intellectual (e.g., lingual, textual); reflective communication; ongoing, serious*
distant partners — *cannot readily have collocated, face-to-face interactions on a regular basis*
local partners — *can readily have collocated, face-to-face interactions on a regular basis*

Fig. 2.1 A visual representation of the design space as defined by the majority of current couple technologies (*left*) and a proposed expansion of this design space (*right*)

Having established a high-level description of the new design space configuration, what are the meanings of *local partners* and *deep interpersonal sharing*, and how do these differ from their proposed opposites? By *local partners* we mean couples that live close enough to be physically present with one another on a regular basis. Local partners need not be in the same room at all times or even when they are engaging in mediated communication, though these are certainly valid configurations for local partners. The difference is that partners at a distance do not have the ready option of being physically collocated or copresent, while those who are local do. This is not a strictly operationalized definition, but judging from the lack of a definition for "partners at a distance," it does not need to be. The value of the term local partners lies in its basic ability to suggest that even couples who can regularly carry face-to-face conversations and engage in physical contact may have the need or desire to participate in mediated interaction. At the extremes of the proposed continuum, we can imagine a couple that lives in different time zones for months on end as opposed to a couple that lives in the same house and spends little more than a few hours apart at a time.

There are at least two issues raised when thinking about designing mediated interactions for the collocated. First, do collocated couples *want* to have technologically-mediated interactions? Look no further than partners who regularly call one another on the cell phone, use text messages or emails, and so on to communicate. One HCI researcher shared with me that she and her husband text message instead of simply talking when they are lying next to one another in the same bed. Mediation, then, can be desirable even when partners have the immediate option of having a face-to-face interaction. Indeed, some past studies have stumbled upon the fact that couples enjoy mediated communication even though they are local (Ito 2005; Bales et al. 2011).

Second, even though local partners desire technological mediation, and often seek it out, is it actually a *good* thing, a design state to be sought after? Technological mediation in collocated situations is often looked upon as a distancing mechanism. See, for example, the Forbes article (Danielson 2007) titled "Is Your BlackBerry Ruining Your Sex Life?" If couple technologies are cast as crutches for those that cannot be face-to-face, then what rationale do we have for inserting computers between partners that are local? Couple-centered motivations for local couple technologies will be explored further in the coming sections.

By *deep interpersonal sharing* we mean the ability for partners to engage in communication that has the power to actually change their mutual understanding. As opposed to the highly abstract, often one-bit, largely visceral exchanges supported by abstracted presence systems, deep interpersonal sharing supports grounding processes in communication that can contribute to "mutual knowledge, mutual beliefs, and mutual assumptions" (Clark and Brennan 1991). So, deep interpersonal sharing often involves verbal, written, or similarly complex and nuanced communicative acts. Furthermore, it involves an ongoing dialogue, such that communication acts can be presented and acknowledged by the other, and shared meaning constructed.

Deep interpersonal sharing may also involve reflective activities that occur at the level of the individual and of the relationship, activities that involve what we call *mutual reflection*. While there have been a number of research agendas aimed at reflection—slow technology (Hallnäs and Redström 2001) and reflective design (Sengers et al. 2005) being two examples—these have largely been focused on how technology might help users reflect on the technology itself. When we talk about reflection, we are instead referring to something more akin to the area of personal informatics (Li et al. 2010), wherein technology supports user reflection on the self. There is one key difference, however, in that mutual reflection is about the self *and* the other. Supporting mutual reflection means supporting reflective processes for the self, the other, and the relationship as a whole[4].

The litmus test for whether a communication medium supports deep interpersonal sharing over abstracted presence is whether or not the communication afforded by the technology can lead the couple to break up. In other words, the medium must allow couples to have serious (but not necessarily un-playful) conversations that can move the partners' interpretations of one another and the relationship forward. Deep interpersonal sharing, then, is about dialogic interactions that can carry highly nuanced meanings as constructed by the partners themselves. Abstracted presence systems, in contrast, tend to constrain the meaning of the communicative acts they enable to a much greater degree. The meaning is largely determined by the designer at the time of making as opposed to the users at the time of communicating. Even

---

[4] Throughout the rest of the chapter, I use the terms deep interpersonal sharing and mutual reflection interchangeably, although the former need not imply the latter.

though some studies of abstracted presence systems have identified the ability for users to layer their own meanings on top of minimal communications (Kaye 2006; Chang et al. 2001), the richness is of a categorically different sort that would not pass the litmus test. So, on one extreme end of the proposed spectrum, we might place single-bit communication devices like Feather and Scent (Strong and Gaver 1996) or devices that send messages without any human initiation like CoupleVIBE (Bales et al. 2011). On the other end, we might place technologies that allow for flexible dialogue or encourage face-to-face interactions. For some technologies, the degree of richness depends on how they are used. Consider the spectrum of uses of video chat systems described by Greenberg and Neusteadter in the next chapter as an example; the video link can be used for abstract awareness ("shared living") or for richer communications like verbal disagreements. So, designing for deep interpersonal sharing merely means providing opportunities for grounded interactions.

Imagining Other Couples

Looking back on the character of existing couple technologies, the following assumption seems to be at work: distant partners are most in need of technological mediation because they cannot touch (e.g., Brave and Dahley 1997), share awareness of the mundane (e.g., Lottridge et al. 2009), communicate subtle emotional expressions (e.g., Strong and Gaver 1996), or generally share intimacy on a regular basis (e.g., Vetere et al. 2005). No doubt, these activities are valued by intimate partners and yet difficult if not impossible to achieve across a distance. But one perhaps unintended implication of this assumption is that couples who are local *do not* have difficulties communicating and sharing in these same capacities. Another implication is that these abstract, lightweight, non-verbal exchanges are the *only* types of interactions that couples, either local or distant, would like to engage in. With the three new design spaces identified in Fig. 2.1, we can begin to explore and challenge these implied assumptions. Not only that, since partners at a distance represent a very small slice—as low as 3.8 million partners according to the 2,000 Census data, or roughly 3 %—of the population (Anon 2008), we can open up the domain of couples technologies to a much wider range of users.

The expanded design space presented here introduces local partners and deep interpersonal sharing to HCI's design landscape towards inspiring radically new designs for couples. It can inspire new questions that lead to a deeper understanding of couples. For example, "what are the design needs unique to each quadrant; what do local partners need that distant partners don't, and vice versa?" It may well be that technologies designed to fit one quadrant can serve the needs of another, as we have already seen in some instances (Ito 2005; Bales et al. 2011). Furthermore, "what types of unique interactions do the technologies in each quadrant afford, and how do these impact the couple relationship?" These are questions which will be addressed in part by this research and that must be considered as the field continues to explore this new territory.

While the benefits of this characterization are many, the expanded space is only one of many possible. As such, it does not represent *the* design space, but rather *a* design space. As much as it helps construct, it helps deconstruct by raising additional design leanings and biases and serving to provoke and encourage further inspection of what technology can do for a broader range of couples. As an example, one might ask "why doesn't Netflix fit on the design space as defined above? How might a system like that be characterized as serving couple needs and what similar technologies might be envisioned?" At base, this work is about helping HCI designers imagine *other* couples towards making technologies that are more relevant to more couples in their everyday lives.

In the next sections, we will shift gears a bit to explain how the new design space characterization emerged from interviews with couple experts. These same interviews also informed the design of a Diary Built for Two, a system that is intended to support mutual reflection for local partners.

Exploring Couplehood and Technology with a Diary Built for Two

Our approach to entering the world of technologies for couples follows that of most efforts to date in that we want to *build something novel*. But, unlike most efforts to date, the technological aspect is only part of the agenda. The other, primary goal of this research is of a social nature; we want to *develop an understanding of couple culture*. So, like Lottridge et al. (2009), the technological component is intended as both a working prototype to be iteratively improved upon as well as a cultural probe to provide social insight. Perhaps the best way to describe the methodology is by characterizing it as Design-Based Research (Hoadley 2004), or *DBR*. DBR is an action-oriented methodology borrowed from the learning sciences field that seeks to develop a functional technology while also contributing to knowledge of situated social phenomena. It achieves this through iterative development and deployment of technologies in real-world settings.

Since the first author, Stacy Branham, is conducting a sole DBR project, the narrative continues in the first person; the design directions and decisions reported in the narrative are hers:

In the context of this research, DBR can be conceptualized as the iterative *design*, *deployment*, and *analysis* of technologies in authentic contexts (Fig. 2.2). In collaboration with Tad Hirsch at Intel, I primed the first DBR cycle by interviewing five couples experts to become familiar with the needs and concerns of couples. This led to a design phase wherein I eventually developed the notion of a shared *mutual reflection* journaling system for *local partners* that I call a Diary Built for Two (aDBFT). I then deployed a low-fidelity prototype of the system to ten couples, collecting among other things over 50 hours of audio recorded interviews with the couples. Currently, I am analyzing the data by transcribing

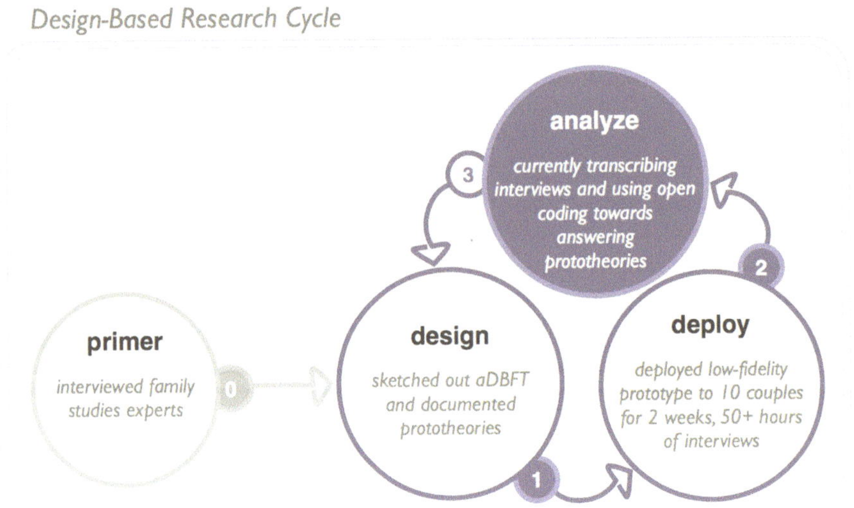

Design-Based Research Cycle

Fig. 2.2 My Design-Based Research approach, through phase 3 of iteration 1

and qualitatively coding the interviews. For more information about DBR, see (Design-Based Research Collective 2003; designbasedresearch.org), but for my purposes here the overview provided in Fig. 2.2 will suffice. In the next sections, I present the primer, design, and deployment phases of the first iteration of my Design-Based Research process.

MFT Perspectives on Couples

As a first foray into the world of couple relationships, I conducted 1-hour, semi-structured phone interviews with five Marriage and Family Therapists (MFTs). MFT is umbrellaed under the larger field of Family Studies, which researches and develops theories around the nature of families and couples. MFTs thus have unique expertise due to their simultaneous exposure to (and sometimes even engagement in) research on couples, as well as their direct experience with couples in therapy—a combination which dovetails nicely with the DBR goal of bridging theory and practice.

Three interviewees were leading MFT researchers at research universities, one was a senior graduate student at a research university, and one was a practicing Licensed Clinical Social Worker. I asked experts about the predominant understandings of couples in the broader Family Studies field. I also asked experts about their experience with couples, including how couples communicate and argue, what needs and concerns couples have, and what role technology has or could have in couple relationships. After transcribing interviews, coding, and discussing themes, a new understanding of couples emerged that suggested providing rich reflective interactions for local partners. In fact, it was the interviews with MFTs that first led to these design considerations, and only after did I find that these design motivations stood in contrast to prevailing couples technologies. A description of how these design considerations emerged from interviews follows, including quotations taken directly from the interviewees.

Connecting Local Partners Perhaps the most interesting idea put forth by the couples' experts is that there is an opportunity for positive intervention within virtually all couples, not just couples seeking therapy. On the one hand, most couples enroll in therapy as a last-ditch effort an average of 7 years after initial symptoms of relationship deterioration arise. In the United States where nearly half of all marriages end in divorce, this means that many couples not currently enrolled in therapy probably should be. On the other hand, every relationship—even a healthy one—can, in the words of one therapist, use a regular "tune-up" and, in the words of another, benefit from "check-ins to remind [the partners] what [they] already know." So, from the perspective of therapists, virtually all couples have needs that could benefit from intervention; from the perspective of technologists, such intervention might possibly be facilitated by interactive devices. This is not to say that technology will replace therapists or even that replacing or mimicking therapists should be the motivational force behind this inquiry. It does, however, raise the question of whether all couples, distant partners being the minority among these, might benefit from therapy-like or therapy-inspired technological mediation.

Therapists also stressed the importance of establishing regular connection between partners. My initial intuitions were proven wrong when the therapists explained that "arguments themselves are not necessarily the problem." Because "the absence of positive in a relationship is more important than the absence of negative," it may be more important to focus on connecting partners instead of trying to curtail arguments. Moreover, "a relationship needs to be rebuilt everyday; it doesn't matter how often you've told somebody you love them, they need to hear it now, or see it now in some form." And, "small deception begets major deception;" chasms between partners may begin with seemingly innocent withholdings about even the most mundane experiences—feelings, daily activities, etc. The key to positive affect, then, is developing patterns of *ritualistic connection*, whether it be through presenting "love gifts"—things like "back rubs, a kind note, a smile, or help in some way"—or by sharing one another's personal feelings and experiences towards developing mutual empathy (e.g., Piercy 2002). Again, these thoughts suggest that from the MFT perspective even couples who are local require daily bids for reconnection—bids that that might be supported by technology.

Supporting Deep Interpersonal Sharing Sharing love gifts may not always be enough to foster positive affect; couples may also need to reflect on themselves and one another in order to connect. For example, in many relationships there is a history of negative "attribution," such that "even the most loving behaviors can be filtered and seen as dastardly, as a negative ulterior motive." One therapist gave the following example: "I did a homework assignment once, kind of a love day, where the person is supposed to say or do things very positively to their partner… This one guy came back and he said 'I told my wife I loved her and she said 'what do you mean by that?' ' " Lightweight acts of connection, perhaps like those enabled via abstracted presence, may require some deeper reflective activities in order to change negative attributions.

Therapists identified that "helping people see themselves differently is a big function of therapy," one that is achieved through "helping couples look at what they're doing." MFTs gave several examples of *reflective* activities that they use with couples to this end. They may ask couples to participate in "meta-communication;" that is, "instead of talking about the contents of the argument," they try to make couples aware of "how [they are] actually having the argument" and how that impacts one another. Another strategy, the "intergenerational approach," helps couples make sense of their interactions by tracing their beliefs and behaviors back to those of their parents. And, "narrative therapy" for couples encourages partners to externalize existing stories about their relationship to co-construct more positive ones (this is called "*re-storying*"). Through these and similar forms of deep interpersonal sharing, therapists enable participants to develop new understandings of themselves that lead to more positive patterns of interaction. Such reflecting and re-storying take place not only at the individual level, but also at the level of the couple in mutual reflection.

Design of a Diary Built for Two

Insights from MFTs led me to explore a range of possible technologies that might support mutual reflection for local partners. Per the suggestion of one therapist, Tad Hirsch and I decided to follow through with the concept of a shared journaling technology. The resulting design, a Diary Built for Two (aDBFT), is a digital personal journal that supports selective sharing between coupled journals (see Fig. 2.3). That is, each partner maintains their own private digital journal, but the system facilitates and hence encourages the sharing of portions of entries between them. aDBFT leans on a diary/journal metaphor towards supporting *ritualized (re)connection* through the *re-storying* afforded by *mutual reflection*.

As depicted in Fig. 2.3, I envision aDBFT running on an interface like the iPad that allows for free-form input from a stylus as well as the option to type entries on a soft keyboard. In its most essential form, aDBFT enables entries to be written and preserved, and allows sections to be shared. Each journal can only be reciprocally linked with exactly one other journal.

The journal/diary genre has important qualities that may be preserved when moving from paper to digital renderings. Diaries provide a private space where one can engage in an ongoing personal dialogue, often on a regular basis (e.g., daily (Mallon 1987)). Additionally, diaries support particularly intimate content; most diaries become grounds for expression of personal thoughts, feelings, and mundane

Re-Pattern   *Reflect / Re-story*   *Reconnect*

Stacy pulls her Diary Built for Two, a mobile touchscreen device, from her bag as she sits down to her daily coffee.

She turns to the next page in her diary and begins to recount and reflect upon her day, an activity that begets new interpretations of herself and her partner.

She highlights as she writes, indicating portions of her journal that she would like to share with her partner.

Fig. 2.3 Design sketch of a Diary Built for Two

experiences that may never be otherwise shared with others (Mallon 1987). As a result of these characteristics, keeping a diary is a highly reflective exercise (Mallon 1987). By enabling the externalization of inner thoughts, the diary invites its author to develop new relationships to those thoughts—whether it be before writing, in the moment of writing, or even minutes, days, or years thereafter.

By digitally coupling diaries and enabling selective sharing between partners, aDBFT may be able to extend the benefits of personal journaling to the level of the couple. aDBFT may support communication between partners at new times of the day or in new forms or even about new topics. I hypothesize that introducing aDBFT will create a space for new patterns of couple interaction—both within and around the system. The framing of the system as being "built for two" and its selective sharing feature may also encourage reflection to move beyond the self to include the partner in mutual reflection. Finally, I hypothesize that the digital diary will reinforce couple (re)connection through ritualistic communication.

While aDBFT's support for mutual reflection is fairly clear, its support for local partners is a little more subtle. Because the sharing feature of aDBFT will likely strip away important context for the sake of privacy, I believe it is necessary that the journal be couched in a communicative space that extends beyond the journal itself. I hypothesize that, for local partners, aDBFT will sit within a shared physical and situational context that will help partners interpret shared snippets. Furthermore, more so than distant partners, local partners will be able to discuss shared portions with one another to elicit missing context. Hence, I see aDBFT as a means and not an end to connection for local partners; the dialogue that begins in the journal does not end there.

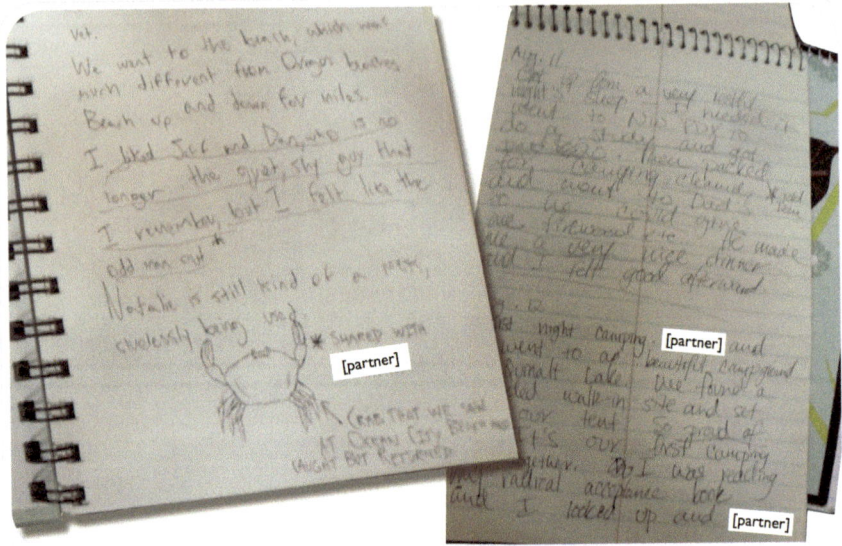

Fig. 2.4 Pictures of P6's and P7's journals. Underlined or starred portions were shared with between partners

Deploying a Diary Built for Two

The hypotheses (or "prototheories," as they are called in DBR) identified above are some of many that I hope to gain insight into as increasingly higher-fidelity prototypes of aDBFT are deployed to couples *in situ*. In addition to addressing the prototheories regarding the technology, I would also like to learn about what it means to be partners in an intimate relationship from grounded examples. This dual purpose along with the fact that the system design was informed by experts instead of couples led me to run a field study with a paper prototype rather than a digital system of aDBFT. With a paper prototype, I could receive basic feedback that might circumvent high-cost technical re-design.

I ran a two-week field study with ten couples. Couples participated in three two-hour interviews and an at-home journaling activity. Interviews were held roughly one week apart at a third space (except for one couple, for which interviews were held at home) with both partners present. Between interviews, participants were asked to keep personal journals using a medium of their choice (a paper journal, private blog, etc.). During the second week, participants were asked to share parts of their journals with one another using a medium of their choice (verbal communication, email, etc.) (Fig. 2.4). The second and third meetings included one-on-one interviews with each partner as well as a joint couple interview.

I recruited participants through craigslist's "volunteering" forum by offering each partner a $50 AmEx card. Replies came from a surprisingly wide variety of couples in a variety of life situations. One couple had been dating for under one year, another had been married for over 20 and had three children together, while another dating couple had children from previous marriages. Some partners were just under 20 years old, others were over 60. Some couples were in precarious economic situations and participated for the money, while others (including a semi-famous author) were financially well-off. One couple was homosexual, the other nine were heterosexual. Couples were not screened in any way—the only requirement being that they self-identified as "a couple"—because exploring the diversity of couple-hood and couple needs is a major goal of this research.

Analysis of the rich data collected in this study is ongoing. The aim is to produce a set of couple case studies that can help guide future design directions of aDBFT as well as guide other designers. Although the analysis specific to how the low-fidelity prototype of aDBFT fared and how its use speaks to the new design space are not yet complete, there are some early take-aways that can be shared. Below, I explore some considerations that seem increasingly important for researchers and designers in the intimate contexts of couple and family relationships.

Managing Values One of the very first issues I encountered in this design-based research was that of value alignment. It was difficult (in fact, impossible) for me to remove myself and my values regarding couplehood from the research. As one very simple example, in considering how to identify potential study participants, I asked the question "what counts as a couple?" My initial thinking did not include polyamorous partners, a thought that my research collaborator immediately questioned: "why not?" The question was an important one which led us to open the study up to any pair of participants that self-identified as "a couple." In this situation, my lack of knowledge about polyamorous relationships led me to unwittingly and without reflection assume that they did not belong to the "couple" category.

Most often, my implicit values did not come to the surface until I encountered an example from a couple (or in the case above, my collaborator) that directly challenged them. When dealing with couples and perhaps families in general, as opposed to dealing with users in an office context, as designers and researchers we are more at risk than ever of bringing personal value "baggage" into the research. For one, we are personally involved in the subject of research; we are members of couples, children of them, members of a culture that is steeped in the couple-hood construct, etc. For another, couplehood is a highly private subject matter that, without more public and open forums for discussion, may be more difficult for us to think critically about. I have come to believe, as others do (Sengers et al. 2005; Allen 2000), that an element of reflexivity may be important in helping us come to terms with our subjective values as researchers and designers in intimate contexts.

Designing for Two Sometimes the most important findings are those that are, in hindsight, completely obvious: when designing for couples, one is designing for two. While my initial concept of aDBFT implied a single interface that would be used by both partners, I am now considering that "one size does *not* fit both." This change of perspective grew out of my realization that each partner had their own identities, perspectives, and needs. As an example, in the interviews with couples,

I quickly noticed that several participants, mostly men, did not identify with the activity of "keeping a diary" in the way that their partners, mostly women, did. However, when I delved deeper, I found that some of the men often engaged in very similar activities of life documentation and reflection. One had kept research and athletic "logs" that documented these activities. But, per the advice given by his research professor and running coach, he also included significant life experiences (e.g., anniversaries and the birth of one's children) in these logs. Another man explained that he reads through the receipts he accrues daily before throwing them away; these provide, he noted, an opportunity for him to reflect on what he has done that day. One design direction I might pursue would entail reframing the aDBFT activity—a task that may be as subtle as calling the "diary" a "journal" or a "log"— to encourage participation from *both* partners.

The heterogenous nature of the couple is an essential problem, one that has been identified in other fields. Tannen (2001), for example, argues that communication between men and women is cross-cultural, and Piercy (2002) accordingly identifies the need for communicative tools that appeal to both partners. The problem of heterogeneity for this particular dyadic user group may be especially important because adoption of new technologies in personal life is more a matter of local consumer choice than in the workplace, where adoption is often enforced by management. For our technologies to be used by couples, they must be designed such that both partners can find a bit of themselves—their values, their interests, their styles, etc.—reflected in them. It may be the case that having symmetric interfaces (as do nearly all of the technologies in Table 2.1) is not the best approach. We need to begin thinking about joint activities/interfaces that can engage both partners, or separate activities/interfaces for each partner that can foster connection while acknowledging individuality.

Power in the Relationship Looking back to the example I gave from my own relationship, it is clear that interactive technologies like Netflix do not enter our intimate lives without stirring up some rather messy interpersonal dynamics. Specifically, Netflix has raised new conversations and stances in my relationship regarding privacy (we had to negotiate a shared password), individual and joint identity, finances, and the subject of this section, power. Who gets to rate a movie? Who gets the benefit of receiving relevant movie recommendations? Who gets to add movies to the instant queue? Who gets treated like the adult? From my perspective, the balance of power is shifted in my husband's favor. The issue of power also came up again and again in interviews with other couples. As just one example, one participant expressed that her husband did not allow her to keep a personal journal at home because he thought it was a waste of time. She expressed to me that she was excited to participate in the study because it gave her the power to keep a journal as she had always wanted to. There is a complex power dynamic in each relationship that will inevitably be shuffled when new interventions are introduced.

Artifacts have long been considered to "have politics," to embed power inequality through the constraints they impose on their use (Winner 1980). The issue of power dynamics is particularly interesting in the case of couple relationships. For many, "inequality begins at home," (Tannen 2001) and some feminist interpretations of power in the home suggest that "the family is the primary site of women's oppression" (Zinn 2000). I believe that designers and researchers for/of couples ought to be concerned with how the interventions they design (this includes researchers' interviews, probes, etc.) may impact power dynamics in the relationship. Not only this, but we should also be aware of the opportunity for us to take a critical/ activist stance, to design our technologies so that they aim to, for example, equalize power within the relationship (Reinharz 1992).

Considering Ethics The final issue I would like to raise is that of ethics. In more than one interview, questions asked of participants led to tense interchanges between partners. In other moments, the questions I asked seemed to bring partners together. In a sense, my interviews mirrored the goals for aDBFT and those presented by therapists; it became a space for reflective conversation between the partners. Rubin and Mitchell (1976) have shown that just the simple act of interviewing couples is itself a therapeutic intervention, one that can lead to long-term strengthening or dissolution of the relationship. On one hand, this presents practical issues regarding evaluation of the actual intervention (in my case, aDBFT), and on the other this means that there is a potential for us to do real harm to a relationship. I discovered midway though the study that several of our participants had mental health issues (e.g., bipolar disorder), which may have made their participation high-risk. And, I may never know if some of the participants were in physically abusive relationships, the violence in which may have been exacerbated by my probing.

In academia the Internal Review Board (IRB) process may help safeguard against certain ethical concerns, but much of the responsibility falls on the shoulders of the researcher—especially when qualitative methods of inquiry require improvisation in situ. In a sense, HCI is entering new territory, whether it be the physical space of the home or the emotional space of the intimate relationship. There may be lessons we can learn from MFT or its parent field, Family Studies, with regard to how to navigate these foreign territories. While we are not therapists nor are we necessarily providing therapeutic devices, we are like MFTs in that we are intervening in relationships. One source we might turn to is the AAMFT code of ethics (aamft.org). As one example, this code requires that "Marriage and family therapists continue therapeutic relationships only so long as it is reasonably clear that clients are benefiting from the relationship." Another accepted method in MFT is to screen participants for mental health conditions and to discontinue the intervention as soon as a participant reveals their condition—even if this means stopping mid-interview. MFTs also have strategies for interviewing that may help guide our efforts, including a list of questions to ask partners before concluding interviews in order to finish the interview in a positive place (Piercy 2002). As designers and researchers for/of couples, we should be considering whether these guidelines from fields with more experience in this space are adequate or fitting within the context of our work in HCI.

Summary

In this chapter, we have explored the technology design space for intimate partners towards expanding notions of what it means to be a couple and what technologies might be built for couples. We have argued that the current design space supports only a narrow set of couples and needs and proposed a new genre of technologies for *local* partners that encourage *deep interpersonal sharing*. And, presenting findings from interviews with couples experts and the design for a device situated in the expanded design space, we have explored some points of caution and contemplation for us as researchers and designers of/for couples. Moreover, our intention here is to encourage ourselves and the community to think more richly about couplehood as a social construct. We need to go outside of our personal knowledge of couples, to go outside of HCI literature, and to go outside of the lab by engaging couples in the field in order to see more and imagine more than we have in the past. In doing so, we may yet generate technologies that are more inclusive and more relevant to couples in their everyday lives.

Acknowledgements Many thanks are in order. Thanks to Intel for funding this work. Thanks also to Tad Hirsch, who was instrumental in the inception of the project, and the many others who have guided it along the way: Joon S. Lee, Clarissa 'K' Stiles, Jason Chong Lee, Deborah Tatar, Dawn Nafus, Christopher M. Hoadley, Fred P. Piercy, Denis Kafura, and Manuel A. Pérez-Quiñones. Finally, we are grateful to the participants who have shared their time, expertise, and personal stories.

References

Allen, K. R. (2000). A conscious and inclusive family studies. *Journal of Marriage and Family, 62*(1), 4–17. [1].

Anon. AAMFT code of ethics. aamft.org. http://www.aamft.org/imis15/content/legal_ethics/code_of_ethics.aspx. Accessed 25 Oct 2011. [2].

Anon. Design based research collective. designbasedresearch.org. http://www.designbasedresearch.org/index.html. Accessed 25 Oct 2011. [3].

Anon. Duofone. gajitz.com. *http://gajitz.com/share-the-love-matching-connectable-phones-for-couples/*. Accessed 25 Oct 2011. [4].

Anon. Hug shirt. cutecircuit.com. http://www.cutecircuit.com/products/thehugshirt/. Accessed 25 Oct 2011. [5].

Anon. netflix.com. http://www.netflix.com. [6].

Anon. (2004). Telesquishy. popgadget.net. http://www.popgadget.net/2004/12/wanna_play_tele.php. Accessed 25 Oct 2011. [7].

Anon. (2005a). Kiss communicator. we-make-money-not-art.com. http://www.we-make-money-not-art.com/archives/2005/10/kiss-communicat.php. Accessed 25 Oct 2011. [8].

Anon. (2005b). Tok Tok & Tug Tug. we-make-money-not-art.com. http://www.we-make-money-not-art.com/archives/2005/01/tok-tok-and-tug.php. Accessed 25 Oct 2011. [9].

Anon. (2008). Love tech goes long distance. forbes.com. http://www.forbes.com/2008/02/06/love-gadgets-valentine-tech-lovebiz08-cx_ag_0206distance.html. Accessed 12 Dec 2011. [10].

Anon. (2009). Daily temperature reading. fatherhoodchannel.com. *http://fatherhoodchannel.com/q2009/12/19/daily-temperature-reading/*. Accessed 25 Oct 2011. [11].

Anon. (2010). Fix a fight. itunes.apple.com. *http://itunes.apple.com/us/app/fix-a-fight/ id376117430?mt=8*. Accessed 25 Oct 2011. [12].

Anon. (2011). Aura. fashioningtech.com. http://www.fashioningtech.com/profiles/blogs/aura-wearable-devices-for. Accessed 25 Oct 2011. [13].

Bales, E., Li, K. A., Griwsold, W. (2011). CoupleVIBE: mobile implicit communication to improve awareness for (long-distance) couples. *Proceedings of the CSCW '11* (pp. 65–74). New York: ACM. [14].

Brave, S., Dahley, A. (1997). inTouch: a medium for haptic interpersonal communication. *Proceedings of the CHI '97 Extended Abstracts* (pp. 363–364). New York: ACM. [15].

Chang, A., Resner, B., Koerner, B., Wang, X. C., Ishii, H. (2001). LumiTouch: an emotional communication device. *Proceedings of the CHI '01 Extended Abstracts* (pp. 313–314). New York: ACM. [16].

Chen, C.-Y., Forlizzi, J., Jennings, P. (2006). ComSlipper: an expressive design to support awareness and availability. *Proceedings of the CHI '06 Extended Abstracts* (pp. 369–374). New York: ACM. [17].

Chung, H., Lee, C.-H. J., Selker, T. (2006). Lover's cups: drinking interfaces as new communication channels. *Proceedings of the CHI '06 Extended Abstracts* (pp. 313–314). New York: ACM. [18].

Clark, H. H., Brennan, S. E. (1991). Grounding in communication. In L. B. Resnick, J. M. Levine, S. D. Teasley (Eds.), *Perspectives on socially shared cognition*. Washington, DC: American Psychological Association. [19].

Clawson, J., Patel, N., Starner, T. (2010). Digital kick in the shin: on-body communication tools for couples trapped in face-to-face group conversations. *Workshop on Ensembles of On-Body Devices, MobileHCI '10*. [20].

Danielson, D. K. (2007). Is your blackberry ruining your sex life? forbes.com. http://www.forbes.com/2007/01/11/leadership-blackberry-treo-cx_pink_0111blackberry.html. Accessed 12 Dec 2011. [21].

Davis, H., Skov, M. B., Stougaard, M., Vetere, F. (2007). Virtual box: supporting mediated family intimacy through virtual and physical play. *Proceedings of the OZCHI '07* (pp. 151–159). New York: ACM. [22].

Design-Based Research Collective. (2003). Design-based research: an emerging paradigm for educational inquiry. *Educational Researcher, 32*(1), 5–8. [23].

Dodge, C. (1997). The bed: a medium for intimate communication. *Proceedings of the CHI '07 Extended Abstracts* (pp. 371–372). New York: ACM. [24].

Goodman, E., Misilim, M. (2003). The sensing beds. *Workshop on Intimate Computing, UbiComp '03*. [25].

Grivas, K. (2006). Digital selves: devices for intimate communications between homes. *Personal and Ubiquitous Computing, 10*(2–3), 66–76. [26].

Hallnäs, L., Redström, J. (2001). Slow technology—designing for reflection. *Personal and Ubiquitous Computing, 5*(3), 201–212. [27].

Hayashi, T., Agamanolis, S., Karau, M. (2008). Mutsugoto: a body-drawing communicator for distant partners. In *Proceedings of the SIGGRAPH '08 posters* (pp. 91:1–91:1). New York: ACM. [28].

Hoadley, C. M. (2004). Methodological alignment in design-based research. *Educational Psychologist, 39*(4), 203–212. [29].

Ito, M. (2005). Intimate visual co-presence. *Workshop on Pervasive Image Capture and Sharing, UbiComp '05*. [30].

Kaye, J. 'J.' (2004). Making scents: aromatic output for HCI. *Interactions, 11*, 48–61. [31].

Kaye, J. 'J.' (2006). I just clicked to say I love you: rich evaluations of minimal communication. *Proceedings of the CHI '06* (pp. 363–368). New York: ACM. [32].

Kaye, J. 'J.', Goulding, L. (2004). Intimate objects. *Proceedings of the DIS '04* (pp. 341–344). New York: ACM. [33].

King, J., Forlizzi, J. (2007). Slow messaging: intimate communication for couples living at a distance. *Proceedings of the Designing Pleasurable Products and Interfaces '07* (pp. 451–454). New York: ACM. [34].

Li, I., Forlizzi, J., Dey, A. (2010). Know thyself: monitoring and reflecting on facets of one's life. *Proceedings of the CHI '10* (pp. 4489–4492). New York, ACM. [35].

Lindley, S. E., Harper, R., Sellen, A. (2009). Desiring to be in touch in a changing communications landscape: attitudes of older adults. *Proceedings of the CHI '09* (pp. 1693–1702). New York: ACM. [36].

Lottridge, D., Masson, N., Mackay, W. (2009). Sharing empty moments: design for remote couples. *Proceedings of the CHI '09* (pp. 2329–2338). New York: ACM. [37].

Mallon, T. (1987). *A book of one's own: people and their diaries*. New York: Penguin Books. [38].

Mueller, F. 'F.', Vetere, F., Gibbs, M. R., Kjeldskov, J., Pedell, S., Howard, S. (2005). Hug over a distance. *Proceedings of the CHI '05 Extended Abstracts* (pp. 1673–1676). New York: ACM. [39].

Ogawa, H., Ando, N., Ondera, S. (2005). SmallConnection: designing of tangible communication media over networks. *Proceedings. of the MM '05* (pp. 1073–1074). New York: ACM. [40].

Patel, D., Agamanolis, S. (2003). Habitat: awareness of life rhythms over a distance using networked furniture. *Adjunct Proceedings of the UbiComp '03*. [41].

Piercy, F. P. (2002). Communication questions for couples: a structure to engage the less articulate, less emotionally available partner. *Journal of Couple and Relationship Therapy, 2*(1), 61–65. [42].

Reinharz, S. (1992). *Feminist methods in social research*. New York: Oxford University Press. [43].

Rubin, Z., Mitchell, C. (1976). Couples research as couples counseling: some unintended effects of studying close relationships. *American Psychologist, 31*(1), 17. [44].

Sengers, P., Boehner, K., David, S., Kaye, J.' J.' (2005). Reflective design. *Proceedings of the Critical Computing '05* (pp. 49–58). New York: ACM. [45].

Strong, R., Gaver, B. (1996). Feather, scent and Shaker: supporting simple intimacy. *Proceedings of the CSCW '96* (pp. 29–30). New York: ACM. [46].

Tannen, D. (2001). *You just don't understand: women and men in conversation*. New York: Harper Paperbacks. [47].

Tsujita, H., Siio, I., Tsukada, K. (2007). SyncDecor: appliances for sharing mutual awareness between lovers separated by distance. *Proceedings of the CHI '07 Extended Abstracts* (pp. 2699–2704). New York: ACM. [48].

Vetere, F., Gibbs, M. R., Kjeldskov, J., Howard, S., Mueller, F.' F., Pedell, S., Mecoles, K., Bunyan, M. (2005). Mediating intimacy: designing technologies to support strong-tie relationships. *Proceedings of the CHI '05* (pp. 471–480). New York: ACM. [49].

Wilson, M. (2009). Hey, who ordered 'Gigli'? The New York times. http://www.nytimes.com/2009/03/29/fashion/29netflix.html. Accessed 07 Dec 2011. [50].

Winner, L. (1980). Do artifacts have politics? *Daedalus, 109*(1), 121–136. [51].

Zinn, M. B. (2000). Feminism and family studies for a new century. *The ANNALS of the American Academy of Political and Social Science, 571*(1), 42. [52].

Chapter 3
Shared Living, Experiences, and Intimacy over Video Chat in Long Distance Relationships

Saul Greenberg and Carman Neustaedter

Abstract Many couples live a portion of their lives being separated from each other as part of a long-distance relationship. This includes a large number of dating college students as well as established couples who are geographically-separated because of situational demands such as work. Long distance couples often face challenges in maintaining some semblance of intimacy given the physical distance between them. Traditional media helped here, where they would stay connected by physical letters, telephones, e-mail, texting, and instant messaging. More recently, many couples resort to "hanging out" over the new generation of video chat systems in order to stay connected. We explore this phenomenon by presenting two composite examples of how couples in long distance relationships hang out over video. Each couple is in a unique relationship situation, yet both share increased intimacy over distance by leaving a video link going between their residences for extended periods of time. These episodes involve couples participating in activities that are sometimes shared and sometimes not, where the key component is simply feeling the presence and involvement of the remote partner in day-to-day life.

S. Greenberg (✉)
Department of Computer Science,
University of Calgary,
2500 University Drive NW, Calgary, AB T2N 1N4, Canada
e-mail: saul.greenberg@ucalgary.ca

C. Neustaedter
School of Interactive Arts and Technology,
Simon Fraser University,
102 Avenue 250-13450, V3T 0A3 Surrey, BC, Canada
e-mail: carman_neustaedter@sfu.ca

C. Neustaedter et al. (eds.), *Connecting Families,*
DOI 10.1007/978-1-4471-4192-1_3, © Springer-Verlag London 2013

Introduction

Long distance relationships (LDRs) are a common reality in this day and age. LDRs include not only people who are geographically separated by large distances, but also those who may be geographically close but who live in different residences. Both share similarities in that access for day-to-day communication is limited. LDRs also include couples at different stages of relationships: from recently-introduced dating couples, to established couples including partners and those who are married. There is a rich literature on the nuances of such LDRs, ranging from academic studies to popular culture "how to" sites that offer advice and experiences to couples.

What is perhaps surprising is that LDRs where people live apart for significant periods of time are not exceptional. Consider dating college students, who often live apart in different cities. Some estimates suggest about 75 % of college students have been involved in an LDR, and that from 25–50 % of students are currently involved in an LDR (Stafford and Reske 1990; Stafford 2005). In another study, 43.6 % of university students reported being in a long distance relationship at some point (Rumbough 2001). Established partners may also find themselves in an LDR (Stafford 2005). Work may force married or domestic partners to live apart for a while. For example, this may result from the assignment of one person to a distant work location or a "two-body problem" where partners cannot find work in the same city (Aguila 2009). Certain jobs often require people to live in different places or to travel for long durations, such as in professional athletics, the military, off-shore oil workers, people who do extensive work in the field, one partner attending an educational institute elsewhere, or mariners who are off at sea. Other non-work factors may come into play (Stafford 2005). Incarceration separates people. Separation may be voluntary, such as dual-career and dual-residence couples who choose to live separately due to career demands, desires for autonomy, and/or desires to live geographically close to family. Crisis (such as ailing parents) may force one person to temporarily reside elsewhere. When taken collectively, we see that LDRs are not rarities. Rather, a good percentage of the population is or has been in a significant LDR (Stafford 2005). For some people, LDRs are highly enjoyable for they provide partners with increased degrees of autonomy along with feelings of novelty (Stafford 2005; Stafford, Meroola and Castle 2006).

Couples in LDRs naturally turn to technology as a tool to mediate their relationship over distance. Historically, they have appropriated non-digital communications technologies to do so, including letter writing and phoning. As digital media and interconnectivity became widespread, they then appropriated emailing, texting, and instant messaging. More recently, free video conferencing software and inexpensive webcams have become available. Consequently, we now see couples adopting and using video chat systems like Skype, Google Chat, Apple FaceTime, or iChat. The general question is: how do LDRs use this new video-based medium?

Specifically, this chapter presents how partners in long distance relationships use video chat systems to maintain intimacy in their relationships. In particular, we examine in-depth instances where a video link is used for long durations of time, i.e., where partners "hang out" together over the link. This goes beyond the more simple phone call-like uses of video chat, where we explore how partners integrate video connections as a core part of their communication routine for extended periods of time in order to enhance intimacy.

We conducted interviews with 14 individuals in serious long distance relationships. We explore and detail two composites from these interviews as example couples: a geographically-close relationship between two adjacent cities, and a geographically-far relationship between two countries. As we will see, in both situations, video is used in a very similar manner, despite the difference in distance and varying relationship dynamics generated as a result.

The main message of these two examples—and of our chapter—is that LDR couples leave video links on for long periods of time primarily because it provides them with increased intimacy regardless of the relationship situation. This intimacy stems from an increased feeling of presence and involvement in each other's lives.

We begin by describing related work on long distance relationship maintenance. Next, we outline our interview methodology from which our two composite examples are drawn. Subsequently, we articulate the details of each example relationship and how video is used to maintain intimacy for the partners, as well as deviations of individuals from our composites.

Related Work

In all relationships, people perform actions and participate in activities that help to sustain their desired relationships—what is sometimes called *relationship maintenance strategies* (Stafford and Canary 1991; Canary and Stafford 1994; Stafford 2005). These include strategic activities that people purposely do to help maintain their relationship (e.g., talking politely) as well as routine behaviors that are simply a part of everyday activities (e.g., cleaning dishes) (Canary and Stafford 1994; Dindia and Emmers-Sommers 2006). Some of the most common interactive activities include acting cheerful and polite, talking openly about the relationship, providing assurances that the relationship has a future, expressing one's love through physical acts, and managing conflicts (Canary and Stafford 1994). Branham and Harrison's chapter in this book builds on this literature by exploring how collocated couples can strengthen their relationship through additional acts of reflection and communication. We also see maintenance strategies relate to how one spends his or her time. This most often includes interacting as a couple with other friends or family who support the relationship, and performing one's share of household tasks or chores. Overall, studies have shown that relationships will deteriorate without the use of a combination of the above behaviors and activities to maintain their relationship (Canary and Stafford 1994).

When it comes to LDRs, the same basic relationship maintenance strategies are used, with the exception of "shared tasks" (e.g., cleaning) since it is harder to perform these over distance (Pistole et al. 2010a). Partners also need to invest in the relationship in various additional ways such as traveling, being available for communication, and financially supporting one's partner, if needed (Pistole et al. 2010b).

Researchers sometimes try to gauge *relationship satisfaction*, where measures are commonly based on satisfaction with several attributes such as one's influence in the relationship, sexual activities, one's own leisure time, division of household tasks, time together, finances, and, most importantly, communication (Vangelisti and Huston 1994). One could argue that LDR partners suffer here. They find it harder to communicate, have fewer sexual activities, less time together and so on, simply because they are not able to see and interact in person as often. If correct, this could cause a lower degree of satisfaction in LDRs. This premise is why many believe that proximity and co-residency is necessary for a satisfactory relationship. However, research has challenged the assumption that proximity is necessary (Stafford and Reske 1990; Stafford 2005). LDRs can be satisfactory because people find ways to achieve the previously mentioned relationship behaviors *in spite of* being separated by distance (Stafford and Reske 1990; Stafford 2005). This is not just an academic argument but one also seen in fact: many LDRs flourish in day-to-day life.

In terms of supporting communication within an LDR, digital media—as realized over the Internet and cellular network—is a potential game-changer. In the past, one defining characteristic of an LDR is that communication opportunities are limited (Stafford 2005). Yet the low cost and ubiquity of digital communication tools seemingly lessens this limitation (Dimmick 2000). Traditional digital media—email, chat rooms, instant messaging, cell phone calls, SMS, texting, and social network sites—creates easier and richer ways for LDR partners to communicate not only with each other but with their common social network. Studies have shown that such digital communication media can ease loneliness and increase feelings of closeness (Aguila 2009) and also increase relationship satisfaction, trust, and commitment while lowering jealousy (Dainton and Aylor 2002). Media is now increasingly rich, and multiple channels provide support for a range of communications—assurance, openness, positivity, and discussing social networks (Johnson et al. 2008; Stafford 2005, 2010)—and even intimate activities like cybersex (Rumbough 2001). Novel research prototypes are even being designed to specifically target couples and the need to maintain their relationships over distance. For example, couples can now share melodies over their cell phones (Shirazi et al. 2009), click to say, "I love you" (Kaye 2006), or—at the extreme—engage in physically-based cybersex via robotic sex toys (Rheingold 2005). However, such technologies are not without their challenges. Scheduling times for communication over such channels is not always an easy task (Aguila 2009) and is certainly more problematic than "bumping into" one's partner while at home. Many communication channels are also not very rich when compared to face-to-face situations.

Within the last few years, a new digital medium has entered the scene: video chat systems that run over the Internet. While video has been available earlier, it often required technical knowledge to use and set it up, it was costly if purchased as a robust product, or it was unreliable and low quality if free. The recent generations of Skype (http://www.skype.com) and other video-based instant messengers have changed this: most computer-literate people can install and use it as a reasonably reliable free service.

Our research question asks: Why and how do people in LDRs use video chat systems? How do they use them in ways that go beyond simple phone call-like conversations, particularly those situations where partners use video over extended periods of time? In particular, does the richer communication channel afforded by "always-on" video better support relationship maintenance over distance? The answers to these questions are the focus of our chapter.

Methodology

We conducted semi-structured interviews with 14 individuals (half female) in long distance relationships. In one instance we interviewed both partners from the same couple. Six interviews were conducted over Skype and the remaining eight were performed in person at either of the researchers' offices. All interviewees were in serious relationships that had moved beyond mere dating, where they considered each other as partners (albeit to a varying degree). Thus, they are couples where each partner would certainly consider the other to be "family." Participants' ages varied from 19 years to their mid-30s. The geographical distance between partners also varied heavily. The closest couple lived in the same city. The furthest apart had partners on the other side of the world, where they were separated not only by distance but by large time zone differences of 10–12 h.

Our sampling is targeted, and we make no claim that it represents a snapshot of the general population as a whole. First, our recruitment process favored calls to the University community; thus our sample tended to have one of the partners being an undergraduate or graduate student, a researcher, or a professor, although it also included blue-collar workers. Even so, the occupations of their partners varied quite heavily. Second, we intentionally restricted our LDR recruitment to those who already used video as one of the primary technologies for communicating with their distant partner, preferably where they kept a video link going with their partner for extended periods of time. Third, we wanted people who had established relationships vs. those who had just met and were still in a very tentative stage (e.g., Internet dating). Still, we tried to stay somewhat general, as we did not select for a particular kind of LDR relationship dynamic. This meant that our sample included quite a few different kinds of relationships in terms of their length, commitment, and relationship dynamics.

What we found remarkable with all of these couples was that each, regardless of the relationship dynamics, was able to maintain large degrees of intimacy through their LDR because the video channel afforded unique opportunities to connect the partners' physical locations and created a shared sense of presence between the partners. By intimacy, we mean that couples were able to engage in activities typical to collocated couples (e.g., deep conversation, shared meals, time together at home, varying degrees of cybersex) where the activities made the partners feel emotionally close and additionally connected with each other. In the following sections, we describe the routines of partners in two types of relationships—short and long distance—as examples that highlight this phenomenon. Our examples were selected in order to emphasize both the diversity of couples' relationship situations and the commonality of how they all used video.

Each example presents the relationship of one couple and their communication routines surrounding the use of video, where each couple is an aggregate of several participants. This was necessary as presenting the results from a single couple in detail risks identifying them and breaching ethical guidelines for the research. Naturally, the aggregation that we have done risks "averaging" the details of our participants' relationships and removing any idiosyncrasies. To circumvent this, after presenting the two composite examples, we discuss any notable differences that we saw between participants. Details beyond these composites can also be found in Neustaedter and Greenberg (2012). It is also important to recognize that the example couples we present are not personas (Cooper 1999; Grudin and Pruitt 2002); instead, they are factual details about our participants, despite being aggregates. All quotes were also told directly to us.

Couple One: Connecting Between Cities

Kaitlyn is 25 years old and has been dating her partner, Tyler, aged 26, for nearly 7 years. Currently, Tyler is a software engineer, while Kaitlyn is a graduate student. Kaitlyn and Tyler lived together for about 2 years before Kaitlyn decided she wanted to return to school to pursue a graduate degree. After carefully talking this through and the effect that it would have on their relationship, Kaitlyn decided to move with a mixture of hesitation and excitement; she was excited to pursue more schooling but would miss being around Tyler day-in and day-out, even though they expected to spend major holidays and the summer months together. They also decided that once Kaitlyn had finished school and was able to move back in with Tyler that the two of them would get married.

Kaitlyn now lives approximately a 2-hour drive from Tyler on the east coast of the United States. Kaitlyn and Tyler have been living apart for 6 months and see each other typically once every other weekend, but this depends on how their schedules permit. Because she is a student, Kaitlyn's schedule is somewhat more flexible than Tyler's so she is the one that travels to Tyler's place so that they can be together (although sometimes they meet halfway in a city between them). When her school workload is light, she can usually leave from school early Friday afternoons and beat rush hour traffic on her way out of town to travel to see Tyler. Visits to see Tyler focus on him at the expense of Kaitlyn's other friends and family who also live in the same city as Tyler. Kaitlyn feels this is unfortunate, but when she visits, she really does want to see Tyler the most and her visits are for such a short amount of time (e.g., 2 days on the weekend) that there really isn't time to see other people.

Kaitlyn was quite satisfied with her relationship with Tyler prior to moving, and this has carried over into their long-distance relationship. She feels that because they have been together for so long, they don't need to say much to each other to communicate. They just need to be together.

When Kaitlyn and Tyler are not visiting each other, Skype plays a critical role in maintaining their typical relationship activities. When Kaitlyn first moved, she started using Skype to call Tyler because she didn't want to have to pay for a land-line phone. This use quickly extended to having long video sessions with Tyler where they frequently "hang out" together.

They've developed a routine around this. Each weekday, Tyler arrives home from work between 5 and 5:30 pm. He phones Kaitlyn around 7 pm, as this is usually when she arrives at her home after work. The call is usually just to coordinate getting onto Skype. If she is ready, the two will start a video chat session on it. The phone call beforehand allows both to stay offline in Skype and only come online to video call each other. They do this because Kaitlyn would prefer to stay offline until Tyler is available, as her mother tends to call at inopportune times.

Kaitlyn and Tyler usually keep their video link going for the remainder of their evening until bedtime, about 4 h, to enhance what Kaitlyn calls *"shared living"* even though apart. During this time, they will most often be "doing their own thing" around the house, while occasionally looking at and chatting with each other through the link. Kaitlyn might make herself dinner, eat, clean the house, do laundry, or sit down to watch some television. Tyler, on the other hand, has usually already eaten by the time Kaitlyn gets home so he will be watching television, playing video games, or sometimes even doing some additional work from home.

> Usually he's sitting on the couch and eating some kind of snack and catching up on, you know, TV…And if there's something that we need to say to each other we'll chime in every now and then…Typically it's a 'we keep it running and live our lives' kind of deal. And it's typical evening stuff, making dinner, making sure things are cleaned up, getting things ready, taking care of personal business, stuff like that. We use video as a method to simulate shared living. Even if we aren't talking, the video channel is open…We do the things we would normally do if we were together and can see one another doing it.

As the quote shows, Tyler sets his laptop on a coffee table in front of his couch so that Kaitlyn can see him most of the time; she doesn't watch him constantly, but will occasionally glance at the Skype window to see what Tyler is doing. Kaitlyn will typically move her laptop between the kitchen and living room, depending on where she is, to keep him in sight. Later in the evening, once she gets tired, she will tell Tyler that she is about to go to bed and the two will end the Skype session.

They also show off new things that have happened to them. For example, when Tyler gets a haircut, he shows it to her. Kaitlyn also shows off the new things she has bought, like clothes and new glasses.

Their routine is fairly static and they will do it day-in and day-out. They love spending time together and the video link provides them with an important opportunity to do this over distance.

Kaitlyn and Tyler also use Skype for conversations more akin to phone calls. However, they stress that it is not just a phone call.

Its really hard to know over the phone to know what's happening in your partner's life. For those reasons seeing someone's body language… its easier to get in there and be closer. … The voice is not enough. The relationship is so physical and visual. Its not just about hearing and talking.

When they do talk, both find it important to be able to see each other, to see each other's reactions, to get a sense of how they are generally feeling, whether they are tired, and so on.

If you asked 'how was your day' over the phone its pretty uneventful. Like if you do it on Skype and actually see the body language the expressions and all that it's pretty good.

Both comment that Skype adds a dimension of empathy not available on the phone, as they can tell how a person is doing from their appearance, facial expressions, and body language. As Tyler says:

I think it just comes down to seeing the person's eyes and smile … sometimes I see her in pretty rough shape on Skype, terrible, like she didn't sleep for a couple of days, overworked, and almost depressed… Its definitely something I cannot catch by phone. I just won't realize what she is going through or whatever, and she'll tell you 'I'm really tired' and all that, but what does that mean? But when I see her like that… her crazy hair and the crazy eyes, well, you can try to be more understanding… at least you know about it. I can do a bit more about it to help, or to say something encouraging.

For them, video also removes a lot of misunderstandings that might otherwise occur over the phone because they can now see each other's facial expressions. Tyler comments:

I always apparently sound pretty harsh when I'm talking or kinda like even when I'm joking it doesn't sound like I'm joking…I would sometimes upset her [on the phone] without even knowing I upset her and of course without intending…With video the problem I had on the phone goes away because she can see that I'm smiling, she can see that I'm being supportive, she can see that I'm not frowning or being angry at her, so you know in that kind of sense it removed those obstacles for us.

Conversations between Kaitlyn and Tyler will happen when the need arises and more often than not they will happen at the onset of their evening together, or just before Kaitlyn heads to bed. Here they both sit down in front of the video link much like a collocated couple might sit down at a kitchen table together to talk. Kaitlyn and Tyler will discuss their day-to-day activities, their biggest worries, plans for seeing each other, and sometimes they will even complain about things or argue. In fact, when they argue, they prefer to do it over Skype so they can see the other person's facial expressions.

Even when we fight we prefer to fight online and see each other because we can see the facial expression of the other person…I think in some cases it can make it worse. In some cases, it can soften it, depending on our reactions really. If say I get so upset I'm bursting into tears, he calms down. Or if something is happening and I'm getting really angry and I'm just ignoring him, he gets more angry so really it depends on the reactions of the person. But the good thing about it is you can see the other person's facial expression because it gives you an idea of what the person is feeling at that moment. If we want to hurt each other more we can, if we want to calm down more we can. It gives us that ability.

Kaitlyn and Tyler also share experiences, such as dinner and television. On some occasions, Kaitlyn and Tyler will spend their time together by having shared dinners, where they plan to both have the same meal and sit down together while they eat. In these cases, Tyler will delay his normal eating time so he can eat with Kaitlyn. Here Kaitlyn and Tyler do not think of their dinner as a video "date"; to them, it is just a normal evening together, much like a couple living together might spend the evening at home together.

> We started having dinner, which has been nice…it'll be a sushi night and we'll get sushi and ahh, umm, so yah, as much as we can to sort of normalize this ridiculous long distance relationship we try… In a way we both know that it's not a date, it's just we're having dinner together in front of Skype. Because it's not a date and I think we're just so used it being casual.

Both like to watch a lot of television and their favorite shows are reality TV ones. Occasionally, they will both plan to watch a show together because they love to see each other's reaction to the sometimes "over the top" antics of the contestants. They also tend to talk a lot as the show airs, and they both enjoy hearing each other's commentary. What makes this routine work well is that they are in the same time zone so the television shows are available at the same time for both of them.

> The reason why we watch together is to see and hear each other's reactions for the shows that we like so much…When we were in [living] together, it was like constant conversation and making jokes and laughing about stupid things people say…it's more like a tool to get to know each other.

While Kaitlyn and Tyler consider everything they do over the link as being intimate, they also do more explicit intimate acts via video. They often 'touch' and 'hug' each other, usually when they have eye contact. Tyler touches by moving his hand close to the camera and doing a stroking gesture (as if touching the other person's face). Kaitlyn hugs Tyler by wrapping her arms around her body in an embrace, and Tyler typically returns the gesture. Kaitlyn will routinely blow kisses to Tyler, especially before falling asleep. He similarly blows them back, but finds it more funny than serious.

They have also tried "cybersex" over the video link but found it less than satisfying. Both found cybersex over video awkward. In spite of being sexually active when physically together, both felt shy in having the other person "watch them." They have now agreed to save their sexual activities for the times they are able to meet up in person. Yet Kaitlyn still occasionally flirts with Tyler to try and entice him for their next visit. Here she will partially unclothe herself and show Tyler, and Tyler would respond with a smile, or a kiss, or a hug.

> But I did like to just strip tease and have this fun with the video and just showing parts of clothing or parts of skin. Like playing with the frame… I'd step away and just show my bra…or showing my back so not really showing everything but still teasing.

Taken together, we can see from the above case that a video link plays a critical role in allowing Kaitlyn and Tyler to share time together when they are apart. The link is about shared living, shared experiences, and shared intimacy.

Leaving the video link open means that they can share an evening together just like they normally do when visiting each other, and like they did before Kaitlyn moved away. It is the presence of each other for these activities that is most important for the two of them, and it is the closest they can get to their normal evening routine while apart. They certainly also use other technologies to connect like text messaging and email, but they are not able to share their time together or feel the other person's presence with these tools. Thus, the video link provides an increased feeling of intimacy between the two partners simply by allowing them to share time together. They stress video is a major contributor to their success. When asked what would happen if Skype wasn't available, they said:

> It would have a big effect. You lose that intimacy. … It's definitely intimacy, all those small things. That's basically all [Skype] is about. And if you don't have Skype, it would be a big deal.

Technical Issues

Despite their successes, there are lots of opportunities for systems to be designed to support their activities better. When directly conversing, mutual eye contact and gaze is certainly challenging for Kaitlyn and Tyler. They also routinely face audio problems. When they are watching TV together, Skype sometimes picks up the sound coming from both of their TVs in addition to the sound of Kaitlyn and Tyler's voices. This makes it difficult to hear and can duplicate the TV show's sound. They resolve this by carefully placing their laptops such that they are far enough away from the TV, but still close enough to them. They could also mute their microphones, however, this would have the negative affect of not allowing them to hear each other's reactions to the show. Lighting can also be an issue depending on where Kaitlyn and Tyler place their laptops in the home (e.g., a dimly lit living room is nowhere near as bright as a well-lit office). Sometimes moving the laptop can be quite challenging, given its weight and the wear of the battery (and its inability to last a long time), and the (lack of) space where Kaitlyn needs to set it in some rooms (e.g., small counters in the kitchen). The connection sometimes fails, or the video quality degrades due to Internet load. However, despite these challenges, Skype allows Kaitlyn and Tyler to do things together that would not be possible without the video link.

Couple Two: Connecting Between Countries

May-ling is 31 years old and lives in a major metropolitan city in Canada. Her boyfriend, Ming, aged 34, lives in China and works at a marketing company where he often works from home. May-ling met Ming 5 years ago when she was living in China. About 2 years into their friendship, she started dating Ming. Several months after this, May-ling received a job offer in Canada as an architect. This was a good career move so she took it. She moved to Canada and continued to date Ming. About 2 years into their LDR, May-ling and Ming were engaged to be married. They plan to get married within the next year and Ming is actively looking for a job in the same city as May-ling in Canada. Once he has work, he will move to be with her.

May-ling and Ming see each other in person only twice a year. May-ling has family in China and so it makes sense for her to visit Ming there; he has no other relations in Canada. May-ling typically travels to China over the "Christmas holiday" break and then once in the summer time when she takes vacation days. She will spend 2 weeks with Ming, but a small portion of this time is also shared with May-ling's parents who live in a city that is a short 2-hour drive away from Ming's home. May-ling really enjoys visiting Ming in person, however, because they only see each other twice a year, the time they do spend together can be overwhelming. They simply aren't used to being physically around each other day-in and day-out.

Their use of a video-based system such as Skype started during this separation out of necessity. While Ming had a webcam, May-ling didn't. Nor had she used Skype regularly.

> It was 2 days after I [arrived] here, and I didn't have a camera. In those 2 days it was very difficult for me. Although we spoke by cell phone and home telephone, it was very difficult for me not seeing him. So I [went] and bought a camera, a web cam. … he already had one, but I didn't.

When they are apart, May-ling and Ming make heavy use of text messaging. They exchange messages sporadically throughout the day, such as good morning greetings, "I love you" notes, and short answers to questions. When they need to have more detailed conversations or to just see one another, they would call each other over Skype. This happens at both work and home. May-ling and Ming talk about their day-to-day activities and the video feed helps to show the other person, which moves it beyond a phone-call like conversation.

> [Video] just makes talking more pleasant and you can see facial expressions. I think that's a really important thing that you miss when you're chatting or talking on the phone…I could not stand not seeing [him]. I mean, I needed him, I needed to see him, and actually everyday we also talk by our cell phone but its not enough for us. I need to see his face. And he also has the same feeling.

As with Kaitlin and Tyler, intimacy and empathy matters.

> We used very lovely words to each other. I always expressed/stated to him that 'I really missed you here' whenever for example I see my friends with their boyfriends or their husbands, 'I really feel you and I feel that I need you to be here with me'…. [We would talk about] how we remembered our past times together, like 'Do you remember when we were at … or when you came home I did this for you'. We do a lot of kissing… And he also used a lot of lovely words towards me, actually because his existence really calms me, I mean when I am upset about things or unhappy he used to hug me and be very kind to me… stroking, hugging and kissing me… he tried to do all those things using the video chat I mean.

In addition to these calls, May-ling and Ming connect their home locations for long durations of time using Skype. In contrast to Kaitlyn and Tyler's LDR where they are both in the same time zone, May-ling and Ming live 12 h apart. This dramatic time zone difference plays a large role in how and when May-ling and Ming connect. Even with such a large time difference, they manage to find a way to "hang out" and video directly supports it. In fact, May-ling estimates that about 80 % of the time, their use of Skype follows the routine described below.

May-ling gets home from work around 6:30 pm, which is 8:30 am in China for Ming and about the time he starts work in the morning. On most days, Ming works from home. Ming knows when May-ling usually arrives home and will send a text message to her around this time to ensure she has arrived home safely. Once he knows she is there, he will call her on Skype. They initiate a video chat session and will then leave it going for the next few hours until May-ling goes to bed. During this time, each continues on with their normal routine. May-ling will cook herself dinner, tidy up the house, read a book, and then get ready for bed. Ming, on the other hand, continues along with his normal work, with the addition that he gets to see May-ling from time to time over the video link. This routine has happened nearly every weekday for the past 2 years. On weekends, their schedules are not normally as routine so they might or might not connect in this way; it depends if both happen to be at home.

While connected, May-ling will move her laptop around the house depending on what room she is in. This includes the living room, kitchen, bedroom, and even bathroom—when she takes off her makeup, brushes her teeth, and gets ready for bed. Sometimes Ming will even see her getting out of the shower after a workout, but this is just "normal" to them and not sexual in nature. Because Ming works from home (and lives alone), there is nobody else around who might happen to see the video link—and thus May-ling—in these compromising situations.

Ming runs Skype on his work computer that sits on a desk at the edge of his living room. Because it is tethered, he cannot move it around the house. He basically sits in front of Skype for most of the time while connected to May-ling. If Ming gets up from his desk, or ventures to the kitchen, he often rotates the camera to the direction of his new location. Ming also needs to regulate the volume and what is visible on his screen though to match his mixed-context of work and personal life. Normally Ming dedicates a small corner of his display to May-ling's video and the rest to his work activities. If he has clients come over for meetings, he must mute the volume on his computer and also hide the video window. For example, May-ling describes one particular instance of being connected to Ming:

> Last night I was watching something on TV and he had a meeting and uh he just cut my voice… I could see him and of course the person he was meeting with couldn't see me but I was just, you know, doing my own thing and no sound but we could see each other… his office is in his house. I was minimized so the person with him couldn't see what was happening on the computer. I just look at him once in a while and then he comes back and tells me he is done and I shush him because I am still watching TV.

Once it is bedtime for May-ling, she will move her laptop to her bedroom so that Ming can watch her fall asleep. This is comforting for both of them.

… I will move [the laptop] to my bedroom, the light is on normally because if I don't turn it on he can't see me…and he normally cuts his voice off so I don't wake up from his phone calls or him talking to people. And at a point in time the computer goes to sleep so it cuts it off….it's on the bedside table and I normally position it towards my face.

May-ling and Ming haven't tried using the video link for sexual acts, beyond just kissing. Their view is that the video link does not provide any real form of physical connection. That is, they consider any acts to be solitary explorations and the video link simply provides a view of the other person doing them. May-ling equates this to a pornographic video without any true connection to Ming.

I've never really had any kind of desire to do virtual sex or anything like that and neither has he, I think… Maybe it's like I'm being watched or something. A lot of times when people ask 'do you have intimate stuff going on online,' I always think to myself that they are talking about a porn movie. I don't want to be in a porn movie for my fiancée.

Taken together, we see that the video link provides an increased feeling of intimacy between the two partners simply by allowing them to have a common sense of "place" and togetherness. Intimacy is not about performing sexual activities together; it is about shared presence. The large geographic and time zone difference means that it is more difficult to participate in shared activities. That is, we don't see May-ling and Ming having dinner together or watching a television show like the first example couple. Their different time zones and schedules don't really permit such activities. Yet this is not a problem because they can still be a part of each other's lives because of the video link. None of the other technologies that the couple has tried have provided such a rich connection for them.

Technical Issues

Like the first couple, May-ling and Ming also face challenges because of the design of the video software and camera. Lighting again is an issue, in particular when May-ling brings her laptop into her bedroom to fall asleep: she needs darkness to fall asleep but Ming needs light to see her. Currently, May-ling compromises. The camera must also be carefully angled in order to capture May-Ling in bed. They use a bedside table but it must be positioned in the correct location, which is not where it normally would be. They've tried placing the laptop right on the bed; however, this made it exceptionally hot and prone to falling over. Audio is again a challenge. In this case, it is because a "work" location transmits to a home location, and the audio must be muted periodically to avoid interruptions and manage the coming and going of work colleagues or clients. Tethering is also a problem: The fixed nature of the desktop computer, power and Internet connectivity, and distance limits of the microphone pickup can all anchor people to a specific location, so they cannot move around the home easily. Certainly, all of these challenges again present design opportunities.

Discussion

Our chapter illustrates how couples in LDRs increase intimacy and maintain their relationships by keeping a video link open for an extended period of time. This creates a shared sense of presence for the couple, even when physically apart. In all of our couples, video enhanced the couple's feelings of shared living, shared experiences, and shared intimacy.

The two composite examples describe the core routines and communication patterns that participants in our study told us about. Certainly we cannot characterize every couple within two cases and, indeed, we saw some idiosyncratic differences emerge between couples. For example, some people preferred different shared activities than the television watching that we presented in the first example. Other couples would listen to music, browse the web or read together (each their own book, but it was still the same activity). Although all our couples had well-defined routines for seeing each other, the frequency and duration of the video connections varied. Some participants would connect every night with their partners, while others would connect several times per week. Nearly all would connect for periods of longer than an hour and most would stay connected from the time they arrived at home after work until bedtime. A small number of participants expressed discomforts about how they looked on the video link, yet the majority did not care about their appearance on camera.

Most couples did have some degree of cybersex, ranging from kissing to nakedness, to flirting, to embracing, to masturbation. However, most did not go that far: one male-male couple reported actively engaging in regular cybersex, while another male-female couple had done it only occasionally. What was common to all our couples was that they described sex—no matter how far they took it—entirely as an extension of intimacy. That is, it wasn't so much about the sex, but rather about being together and being intimate together.

Still, nearly all couples expressed similar issues of "awkwardness" in regards to performing hard-core sexual acts over the video link. This ranged from some feeling that it was somehow "wrong," to others just not finding it that satisfying, to others that didn't pursue it because they were concerned that the video channel wasn't secure, i.e., that an outsider could eavesdrop and even record their sexual act.

There were also participants in our study who fell somewhere in the middle of the two example couples in terms of their geographical distance apart. The first example explores couples who are in the same time zone and a few hours drive apart, while the second example looks at connecting across many time zones. A number of our participants were somewhere in-between these ranges, where they were apart by two to three time zones across continental North America. Even in this seemingly small time-zone difference, the difference was still enough to affect the couples' routine. In these situations, shared meal times were not possible. Yet people did find a way to develop routines. Most couples could still connect, and most often did, during the evening. For one partner it was early evening and for the other it was the late evening. This sometimes meant adjusting one's sleep cycle to accommodate the need to have shared "together time." A more broad discussion of time-zone challenges for family communication can be found in Cao's chapter within this book.

Overall, our interviews and composite couples reveal a pattern of communication that has moved beyond phone call-like usage. Even when couples conversed, the video added a crucial element of seeing the other person's face and facial expressions. Even more radically, couples have appropriated video technologies in a new way that makes more sense to them: They have turned video chat systems into tools that connect two locations in a more permanent fashion. It isn't so much about conversing as it is about shared living. This usage begins to look dramatically similar to media space systems of the 1980s and onwards that saw industrial research labs and universities (e.g., PARC, EuroPARC, University of Toronto) connect distributed offices, workspaces, and buildings with "always-on" video (Harrison 2009). We also see this theme emerge more broadly in this book; Judge, Neustaedter, and Harrison's chapter reveals how families with children also find value in leaving their video link open for an extended period to connect with grandparents or sibling families.

Yet video as used by LDRs is much more than sharing a living space with a colleague: significantly, LDRs appropriate the channel as a way to maintain their intimacy and their relationship. This was successful for our participants because all shared a relationship (to varying extents) prior to moving apart. It is possible that long distance relationships formed *over* a mediated link would exhibit different behaviors as research on workplace media spaces has shown that media space systems are better at sustaining existing relationships than helping to initiate new ones (Harrison et al. 1997). Couples in a long distance relationship may also very rarely have face-to-face encounters. When distance separation is extreme, such as for Ming and May-ling, the relationship may be nearly entirely mediated by the video link. This could easily create challenges when the couple reunites in person and was indeed the case for several of our participants.

There are also other issues that make using a video link for extended periods of time challenging. Pragmatically, it can be difficult to situate and move a computer, even if it is a laptop, to the various locations that one may wish to broadcast his or her life from to the remote partner. There are also problems related to camera angle, lighting, and audio. While not discussed in our two examples, many participants similarly told us that it was sometimes difficult to keep their video connection going for longer periods of time because of software issues with their video chat system and because of variable performance of the Internet. These are all technical issues that need to be addressed through design and implementation.

In addition, many social issues exist that are perhaps more difficult to solve through design. People are hesitant to broadcast video for extended periods of time from work or they may not be allowed to; this forces connections into the evening hours. Sometimes people can work around this by working from home. Yet this brings challenges with connecting mixed contexts, namely work and home, as seen in the second example. There are also challenges in moving from shared time together to intimate sexual activities. Currently it is not possible to truly connect with a remote partner in a physical sense when using a video connection because the technology is lacking. Video chat systems are simply not designed with cybersex in mind, akin to the way that sex toys are now being carefully designed for aesthetics, embodied pleasure, and intimate experiences (Bardzell and Bardzell 2011). This turns a design problem into a social issue where feelings of awkwardness or embarrassment arise when couples try to use a video chat system for sex acts, but are unable to do so.

Conclusion

Our chapter has explored the ways in which couples in long-distance relationships stay connected by using a video chat system. In particular, we have focused on describing how couples increase intimacy by leaving a video link open for an extended period of time. This has opened up the possibility for couples to share a variety of activities together while apart. It has also enabled couples to connect their residences together such that they can continue on with their normal routines, only now a remote partner can see and even be a part of them in a way that is not possible with other technologies. This suggests an avenue of design that directly supports creating a shared sense of presence between partners in long-distance relationships. This should certainly include systems that utilize a video link, but they may also include other mediums. The crux is finding and utilizing mediums that provide a rich enough experience that partners feel they are actually a part of their remote companion's life.

References

Aguila, A. P. N. (2009). Living long distance relationships through computer-mediated communication. *Social Science Diliman, 5*(1–2), 83–106.

Bardzell, J., Bardzell, S. (2011). "Pleasure is Your Birthright": digitally enabled designer sex toys as a case of third-wave HCI. *Proceedings of the CHI* (pp. 257–266). New York: ACM.

Canary, D., Stafford, L. (1994). Maintaining relationships through strategic and routine interactions. In D. Canary & L. Stafford (Eds.), *Communication and relational maintenance* (pp. 3–22). San Diego: Academic.

Cooper, A. (1999). *The inmates are running the asylum*. Indiana: SAM.

Dainton, M., Aylor, B. (2002). Patterns of communication channel use in the maintenance of long distance relationships. *Communication Research Reports, 19*, 118–129.

Dimmick, J., Kline, S., Stafford, L. (2000). The gratification niches of personal email and the telephone: competition, displacement, and complementarity. *Communication Research, 27*, 227–248.

Dindia, K., Emmers-Sommer, T. M. (2006). What partners do to maintain their close relationships. *Close relationships: functions, forms and processes* (pp. 305–324). New York: Psychology Press.

Grudin, J., & Pruitt, J. (2002). Personas, participatory design and product development: an infrastructure for engagement. *Proceedings of the Participatory Design Conference*.

Harrison, S. (2009) *Media space: 20+years of mediated life*. New York: Springer

Harrison, S., Bly, S., Anderson, S., Minneman, S. (1997). The media space. In K. Finn, A. Sellen, S. Wilbur (Eds.), *Video mediated communication* (pp. 273–300). Mahwah: Lawrence Erlbaum

Johnson, A. J., Haigh, M., Becker, J., Craig, E., Wigley, S. (2008). College students' use of relational management strategies in email in long-distance and geographically close relationships. *Journal of Computer Mediated Communication, 13,* 381–404.

Kaye, J. (2006). I just clicked to say I love you: rich evaluations of minimal communication. *Extended Abstracts of Proceedings of the CHI*. New York: ACM.

Neustaedter, C., Greenberg, S. (2012). Intimacy in long distance relationships over video chat. *Proceedings of the CHI*. New York: ACM.

Pistole, M. C., Roberts, A., Chapman, M. L. (2010a). Attachment, relationship maintenance, and stress in long distance and geographically close romantic relationships. *Journal of Social and Personal Relationships, 27*(4), 535–552.

Pistole, M. C., Roberts, A., Mosko, J. (2010b). Commitment predictors: long-distance versus geographically close relationships. *Journal of Counseling and Development, 88,* 146–153.

Rheingold, H. (2005). Teledildonics and beyond. In B. Arthur (Ed.), *The postmodern presence: readings on postmodernism in American Culture and Society* (pp. 274–287). New York: Altamira.

Rumbough, T. (2001). The development and maintenance of interpersonal relationships through computer-mediated communication. *Communication Research Reports, 18*(3), 223–229.

Shirazi, A., Alt, F., Schmidt, A., Sarjanoja, A., Hynninen, L., Hakkila, J., Holleis, P. (2009). Emotion sharing via self-composed melodies on mobile phones. *Proceedings of Mobile HCI*. New York: ACM.

Stafford, L. (2005). *Maintaining long-distance and cross-residential relationships*. Mahwah: Lawrence Erlbaum.

Stafford, L. (2010). Geographic distance and communication during courtship. *Journal of Communication Research, 37*(2), 275–297.

Stafford, L., Canary, D. J. (1991). Maintenance strategies and romantic relationship type, gender, and relational characteristics. *Journal of Social and Personal Relationships, 8,* 217–242.

Stafford, L., Reske, J. (1990). Idealization and communication in long distance premarital relationships. *Journal of Family Relations, 39*(3), 274–279.

Stafford, L., Merolla, A., Castle, J. (2006). When long-distance dating partners become geographically close. *Journal of Social and Personal Relationships, 23*(6), 901–919.

Vangelisti, A., Huston, T. (1994). Maintaining marital satisfaction and love. *Communication and relational maintenance* (pp. 165–186). San Diego: Academic.

Part II
Immediate Families and Children

Chapter 4
Intra-Family Messaging with Family Circles

Ruud Schatorjé and Panos Markopoulos

Abstract This chapter makes the argument that intra-family communication is not an issue of connectivity anytime anywhere, but of providing communication media that are flexible and expressive allowing families to appropriate them and fit their own idiosyncratic ways of communicating with each other. We examine households with working parents and teenage children who are starting to find their own way in life, developing separate routines and social networks outside the family. We found that despite both generations being users of various modern media, opportunities for communication are not always taken and there is a less than desired exchange of expressive and affective messages. We sketch this design space by briefly describing some earlier works. Furthermore we present a reflective account of the design of Family Circles and some lessons learnt from its preliminary evaluation.

Background

Recent years have brought about a steep increase in the availability and use of technologies that support informal and social communication. Following the practically total adoption of email and mobile phones in developed societies, text messaging, blogs, micro-blogs (e.g., Twitter) and online communities (e.g., Facebook, Google+) are growing rapidly in popularity, particularly amongst young people. The often heralded ambition to connect anytime and anywhere is by now in many ways a daily reality to which people are becoming accustomed to and are even starting to expect. Modern living has become saturated with opportunities and means to engage in social communication even to the point where people find it challenging

R. Schatorjé (✉) · P. Markopoulos
Eindhoven University of Technology,
Eindhoven, Netherlands
e-mail: mail@ruudschatorje.nl

P. Markopoulos
e-mail: p.markopoulos@tue.nl

C. Neustaedter et al. (eds.), *Connecting Families,*
DOI 10.1007/978-1-4471-4192-1_4, © Springer-Verlag London 2013

to cope with expectations availability for communication, responsiveness, and disclosure through such media.

Perhaps surprisingly, this abundance of communication media and opportunities for communication does not automatically translate to improved intra-family communication. Research in the Netherlands found that between the mid-seventies and the millennium, direct contact among household members in Dutch households decreased steadily (Breedveld and van den Broek 2006; Breedveld et al. 2001). There are many reasons why this could be so. Breedveld et al. explain that post-millenium family members spend more time at home but less time in conjoint activities (e.g., engaging in conversation, visiting friends and relatives together) compared to earlier years, and describe a growing individualisation within families. Families congregate far less than before at a single central place in the home to watch TV for example. Technology is spread around the home letting household members engage individually with media (de Haan and van den Broek 2000). Starting from early adolescence, parents and children gradually spend less time together and exhibit less physical affection toward one another (Richardson 2004). Even though open communication between young adolescents and their parents has increased, adolescents' perceptions of family cohesion, family satisfaction, and intimacy are declining. Simply put, working parents are often out of the home while teenage children follow their own schedule and are often reluctant to spend free time with the family at home.

We remark that the wealth of media sketched above does not seem to provide the answer. Mainstream communication technologies target primarily people separated by distance rather than those sharing a household. Worse, interaction with remote others takes up time at the expense of time spent with family members limiting the opportunities for in depth and expressive communication. Social media allows people to easily connect to many others, creating many superficial and ephemeral relationships. However, this contact often consists in one or two line-messages, limited in affective expression and multi-cast rather than personally addressed (Lenhart et al. 2007).

In the quest for technologies to address this problem, we take the position that the challenge is not one of quantity but of quality of communication. We present a design based exploration of how technologies might support intra-family communication, compensating for these trends, and complementing existing media. We targeted intact families with teenage children, where the difficulty of aligning routines and spending time together noted above are most prevalent, and we examined how we can support transitory messaging. The design case we present, is aimed at illustrating how further to the interactivity afforded by the digital medium, the form giving, the detailed low level interaction design, and the core functionality of the system are closely knit elements determining the nature of the emerging communication experience.

The remainder of this chapter describes the design of Family Circles, a system designed to support transitory messaging in the household. We start by introducing the design challenge based on related literature, the process of the design and its final product as well as the evaluation of the concept with users in actual life. We conclude reflecting on the topic of intra-family communication and discussing links to related research and design works.

Related Work

There is a substantial body of research that examines intra-family communication. Some often visited scenarios concern video communication between remote family members (Yarosh and Abowd 2011; Judge et al. 2010), technologies to support awareness of remote elderly relatives, e.g., (Mynatt et al. 2001; Dadlani et al. 2010; Metaxas et al. 2007), or supporting the sharing of mundane daily experiences (Markopoulos et al. 2004; Davis et al. 2007). Common to these works is an emphasis on bridging physical distance and countering the goal oriented nature of synchronous communication by telephone or messaging.

Recently there has been an increasing interest to support communication of family members living in the same household. A well known example is the Whereabouts Clock by Microsoft Research, a research prototype that provided location awareness for family members (Brown et al. 2007) and that was shown to support both practical but mostly affective needs of the family even with very coarse location information. Khan et al. report a multi-method research on communication needs of busy parents that are not well supported by existing media (Khan and Markopoulos 2009), that included interpersonal awareness and expressive communication. They went on to design Family Aware, a dedicated mobile application supporting awareness between the two parents thorough (Khan et al. 2010), but which did not go further than simple text messaging with regards to supporting affective communication. Other systems supporting opportunistic and transient text messaging between family members are the HomeNote system (Sellen et al. 2006) and StickySpots (Elliot et al. 2007) which can receive and display mobile text messages as well as locally scribbled messages using a touch screen display.

A more playful approach to this challenge is the Photomirror appliance (Markopoulos et al. 2005) intended to support intra-household communication. The Photomirror supported awareness of commotion in and out of a house by automatically registering departures and arrivals in the hallway as still pictures, but also supported expressive communication through short video clips. All information captured was ephemeral, with a decay time of a few hours, after which it would not be retrievable. Brief field trials suggested that the automated capture of stills for supporting intra-family awareness provided fewer benefits than the explicit intentional capture of video clips which gave rise to playful and engaging exchanges. It appears that for household members, even though they share quite a lot of their daily life together, autonomy and control over information remain important and a great value is placed upon expressive communication. Importantly, Photomirror illustrated how the emerging experience was context sensitive, even fragile, with regard to the location of the home where the device was placed.

Fig. 4.1 Fida can collect and store conversation topics a child wishes to discuss. (Zoontjes 2007)

The sharing of transitory and playful video clips as with the Photomirror has a direct analogue to common every day practices regarding paper written notes. Taylor and Swan examined the location where such notes are left and argued that it varies according to the nature of the message and the intention of the sender (Taylor and Swan 2005). Messages of organisational nature are often left in the kitchen, e.g., on the kitchen table, the counter of the fridge surface. When an item is dealt with, the note is removed from its position and stored at a pile elsewhere so it is clear it has been resolved or a messages is received. Rather than a device bound to a specific location, it appears that intra family messaging requires some flexibility for choosing the location depending on the message.

Two related system concepts developed in our department, that aim primarily to improve the quality of communication between family members, are Fida and Jakob. Fida (Yalvac and Helmes 2007) was a design concept which examines how to lower the barriers for communication between young adolescents and divorced parents. Rather than going the obvious route of increasing availability and ease of use of the technology, the designers targeted perceived barriers for children to initiate discussions with their parents face to face. Fida supports and invites the recording of brief messages that can be shared with parents in a non-confrontational manner and invite discussion at a later instance. Jakob (Kassenaar 2009) is an interactive couch designed for leaving messages for other family members so that they open discussion at a later occasion. It attempts to compensate for the decreasing amount of time people spend talking to each other. The couch is able to play back recorded messages, either the last recorded or a random older message. Hereby it functions as a tool for slow (non-urgent) communication to help keep family members aware of each other's' feelings and activities (Fig. 4.1).

Understanding Family Communication

A study of intra family communication practices and needs was conducted using a combination of information probes and video recorded contextual interviews. As this project focused on families with parents both working a substantial amount of hours per week, this was one of the main requirements for the study's participants. Furthermore, the family was required to have at least two children, both between 12 and 20 years of age. Because of time constraints, only two families could be recruited.

Information Probes

The information probes study aimed to explore the transitory message dynamics present in a household with teenage children and to identify opportunities for this project to enrich the existing messaging system. The probes method was largely based on the information probes described by Hemmings et al. (2002) (this method is a variant of Gaver's Cultural Probes) (Gaver et al. 1999). This method required participants to make a slight change in their normal behaviour concerning the use of scribbled messages left in the home for another person.

The probe deployment lasted 5 days. It started with a briefing session of two parts: a video-recorded semi-structured interview (as described in the next paragraph "Contextual Video Interview") that aimed to acquire information about the specific family and its indirect messaging dynamics; and an explanation part to clarify what is expected of the family during the next 5 days. The debriefing session took place at the end of the 5 days. It was an open discussion about the study and about the subject of the project.

Probe Package Content

The initial probes package consisted of three main items:

- A disposable camera to capture daily experiences at home;
- A booklet providing several questions about the family and their habits and some tasks to do for each day of the study;
- Sticky notes in different colours, one colour for each family member, to be used to their own likings and as indicated in some tasks.

For the second family, the sticky notes were replaced by message-cards and a larger central messaging board. The message cards contained two textboxes to write down a message. The initial message could be written in one half while a response to it could be left in the second box. The message board was a piece of cardboard with several textboxes arranged in a vertical manner. Family members could leave replies to the messages above and by doing so they could form a timeline of indirect communication events. The rationale for this change was to probe more specific reactions about how people use the option to reply to a message and to learn how people respond to having a message board at a central place where family members can leave (perhaps more informal) messages and respond to other's.

Contextual Video Interviews

The video recorded interviews in context, were aimed at learning more about people's motivation to leave messages for others and about the dynamics and contextual nature of the family's indirect communication. They were open, semi-structured interviews guided by pre-determined, questions but more important, by matters that are encountered during the discussion or at places in the home. By zooming in, both literally with the video and metaphorically with the questioning, on the issues relevant to intra family communication, we aimed to get insights and inspiration. The interviews were conducted among two families of four with children between 12 and 18 years of age and took place in the homes of the families. All four members of the family were present on the moment of the interview and answered questions of the interviewer while another person recorded everything relevant on video.

Results

As one might expect, participants indicated that most communication occurs verbally and face to face. Indirect messages are written mostly for practical and organizational purposes (e.g., "Could you turn the dryer on when you read this?") or even as a reminder to oneself ("Don't forget to take bread out of the freezer before going to bed!"). Some messages were not addressed to a specific family member but to the one that gets to them first. Organisational messages are often positioned in a central position in the home. This was usually the kitchen as this was a part of the house that every family member passed through on entering the home. If messages are addressed to one person, the name of that person is mostly written at the top of the message. The second family relies more on using the phone, apart from the mother who is not so comfortable with it and prefers to write down messages. The rest of the family indicated that picking up pen and paper and actually writing a note was too much of an effort and chose to use the phone instead.

The probe study exposed the families' structure and messaging habits. Although both parents of the first family had a full-time job with a lot of responsibility and the children were engaged in several extra-curricular activities, they knew reasonably well what other family members were up. But perhaps more important: they were willing to align their own activities to ensure at least one 'quality' contact moment per day, mostly during dinner. Before the probe study, the family already used scribbled messages to communicate organisational messages (e.g., reminders, requests, tasks or informational messages). When they were encouraged by the probe's booklet to leave a message at a given time, messages were often experienced as more informal and more personal. The position of the messages shifted from the kitchen to a more decentralized place, more directed at the intended receiver's habits (for instance the receiver's bedroom door). As the family also embedded this transi-

Fig. 4.2 The nature of actually written messages varied from the kind of messages that families desired to communicate

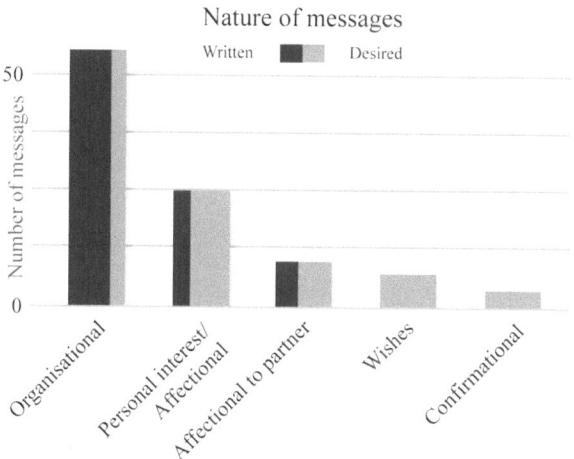

tory messaging into their domestic communication dynamics, they did not experience it as a burden, although they did favour direct communication over leaving scribbled messages. From their perspective, paper notes were somewhat impersonal and lacked feedback as to whether a message is received as intended or whether it is received at all.

The second family was a little less typical. As the parents (47 and 49 years old) owned a restaurant, a lot of their time was devoted to their business. They reported having dinner together as a family 3–4 times per week. Both sons (18 and 20 years old) were in college and were considered old enough to take care for themselves when their parents were not around. They were not so much used to leaving scribbled messages for each other, since they usually would take the phone to contact someone. While they believed that scribbled messages would function well as a reminder, they were afraid that a message is easily missed and considered the phone to be quicker and more effective. This family got to use the message cards that were offering the possibility to write a reply underneath an earlier sent message. However, because of the somewhat chaotic organisation in the home, messages were, as the family already indicated to be a likely event, missed by the intended receiver and it often appeared that the original sender also wrote a message in the reply-area of the card, just to fill up the space. The central message board that was positioned in the kitchen for the study did not become a part of the messaging dynamics of the family. As they were unable to think of something useful to leave on the board, it remained blank until the last day when the father decided to fill-up the empty space because he believed handing it back in unwritten was not polite.

In Fig. 4.2, an overview is displayed of the messages that were written or described in the booklet by both families. The most common kind of messages were those of organisational nature. To this category belong messages that act as a reminder, inform someone, ask for a favour or order someone to do something, in short: messages that assist in the families daily routines (e.g., "Could you turn on

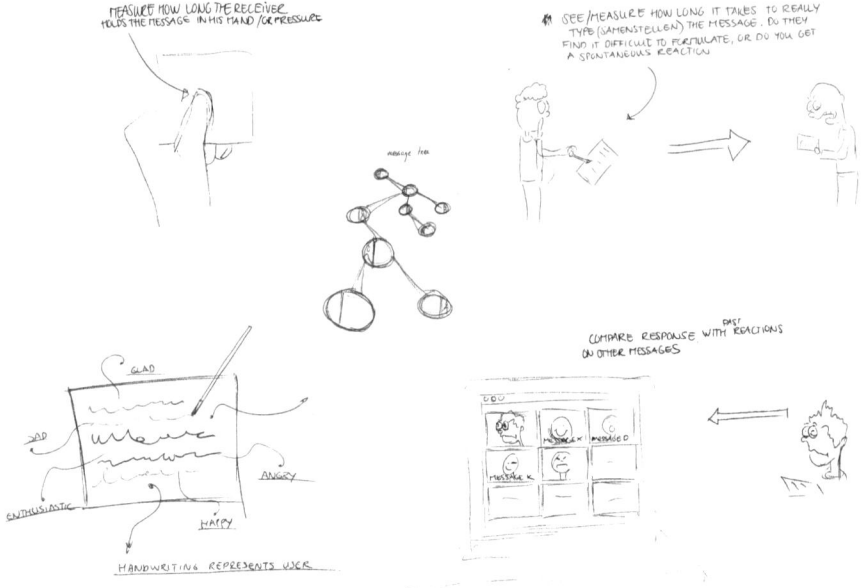

Fig. 4.3 The sketches above display early explorations of domestic indirect messaging concepts. The explorations addressed aspects such as the sender's or receiver's state of mind; (emotional) response to a message; or numerous possibilities made possible by different physical shapes

the dryer when you get home?"). In the second category, named Personal Interest/ Affection are messages that aim to learn more about a family members experiences (e.g., "How was the soccer match?") or sent out a signal of affection (e.g., "I am going to miss you"). Interesting to note here is the difference between messages desired, the right, lighter half of the bar in Fig. 4.3, and messages actually written, the left, darker half of the bar. When the family members were asked about this during the debriefing session, they indicated that even though they liked sending and especially receiving notes of a more affective nature, they had not done so as their plans were to tell in person at a later moment, which they often did not follow through on. The third category is about personal messages to one's partner. This category houses more intimate messages (e.g., "I love you") that show much affection but have less practical use. The fourth category is one mainly housing messages from children. It describes merely fictional messages that have no chance to be taken seriously (e.g., "I hope mum and dad suddenly give me 1,000 € allowance"). The fifth and last category describes messages that ask for reassurance (e.g., "Have you cleaned out the rabbit hutch today as I told you?"). Although participants indicated that messages of the latter two categories to be desirable, in reality they were never created.

Conclusion and Design Implications

Looking at the graph, there thus is an noteworthy difference in the desirability of organisational and affective messages and the actually written amount of these messages. Further, the place where a message is left varies greatly upon the content and nature of the message. Despite the limited magnitude of the study, several directions for the design concept were identified:

- Enhance communication quantity; as described by Noller and Bagi (1985), stimulating the quantity of communication might well have a positive effect on the quality of the communication and the social connectedness between family members.
- Emphasize on presence-in-absence. As earlier research has shown (Ijsselsteijn et al. 2003), being reminded of the ones close to you at their absence enhances the feeling of connectedness with those persons.
- Enable more personal communication. During the probe studies and contextual video interviews, participants repeatedly stressed the impersonality of the used post-it notes. Therefore the project should aim to intensify or sustain the feeling of being-in-touch and to enable more personal messages with an increased the emotional value.
- Stimulate a conversation, instead of 'one-way' commenting. As indicated by an expert of the Dutch 'Centrum voor jeugd en gezin' and implied by the work of Lenhart et al. (2007), de Haan and van den Broek (2000) and Richardson (2004) (among others), stimulating two (or more) way-communication for which the overall trend is declining, to become an essential part of this project. It should attempt to encourage every individual of the household to take part in the communication.

Already at this stage in the design process, sketches were made in a creative session with four industrial design students to explore different directions of indirect messaging, as shown in Fig. 4.3. Ideas were simple yet numerous and allowed us to broaden the scope of the project.

Technology Probe Study

We decided to focus on voice messaging. Voice messages are easy to capture, they can be expressive and they can be displayed (played) at different locations in the home easily. This is important for addressing recipients privately and in a personal way. A technology probe was created that would support voice messaging and would let us see how a simple messaging functionality is appropriated by a household. This study was largely based on Hutchinson's Technology Probes (Hutchinson et al. 2003) but unlike Hutchinson, we did not collect the data in such a structured way. Our technology probes solely served the purpose of revealing new

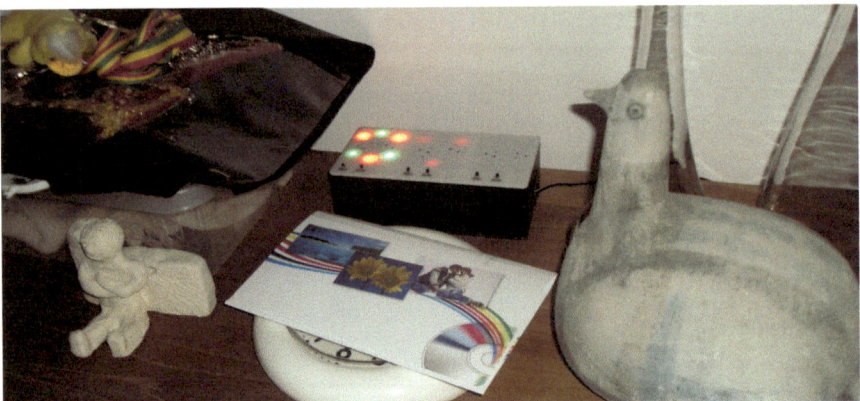

Fig. 4.4 The prototype was placed in a central position in the families' home, this particular family chose to place it on dresser in the living room

design opportunities by triggering and enabling reactions by participants and observing how the introduction of technology could change behaviour and communication patterns. The probe consisted of three voice messaging slots in a single object (see Fig. 4.4 middle). Each message slot was adorned with secondary information using red and green light signals.

The initial rationale behind the three messaging slots was that a reply to an earlier recorded message could be recorded at the slot next to it. This would eventually create a short thread of messages. The red light in this technology probe study communicated which message was recorded last and also tried to provoke the users to record a reply by pulsating red light in the slot next to the one that has just been played. The red light indicating new messages faded out more after each time that message had been played, and disappeared completely after three plays, indicating that the message was not new anymore. The green light in the picture visualizes how many times a message has been played (brighter is more), visualising thus the popularity of a message in comparison with the other two slots.

Results

The technology probe was deployed with a single family of four with both parents working full-time and teenage children. The family was not instructed to record their use of it, nor were they asked to execute any specific tasks for the study. All messages were stored in the device for retrieval after the deployment period (Fig. 4.4).

During the debriefing the mother indicated that at first use, the family (mainly the children) used the device for the most part for funny, light-hearted voice recordings but after getting used to having such a device at hand, messages began to take a more functional nature and became aligned with the existing transitory messag-

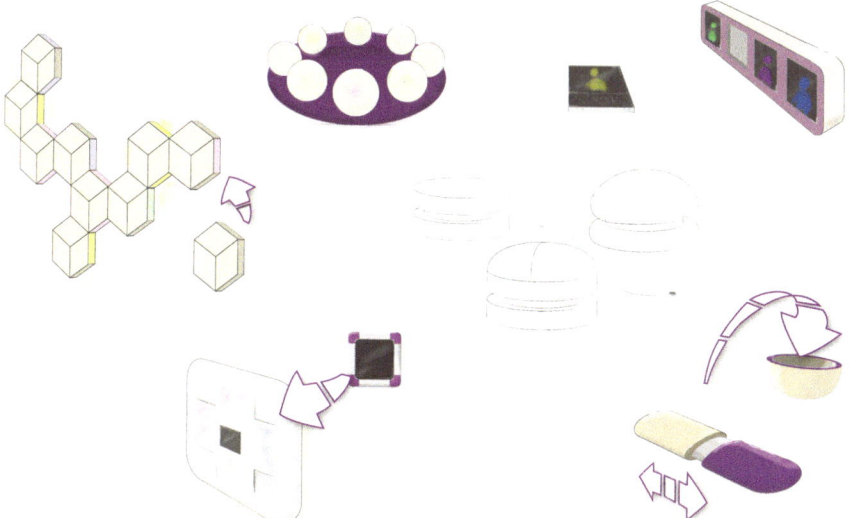

Fig. 4.5 The sketches above display explorations of various physical shapes and how they can be used and stored in the home

ing dynamics of the family. With the family's growing experience with the device they deviated from the linear construction of message threads that the probe was designed to support, recording one message over another without paying attention to which slot contains the oldest message. Overall, the children in the family used it more than the parents. Furthermore, the coloured lights seemed to prompt the family to use the device. Although the children in the family used the system spontaneously from the moment of installation on, the parents indicated to need the lights in the device mostly persuaded them to use it. This was something that was already encountered during the creation of the working prototype that was used in this exploration. People that passed by were tempted to leave simple, funny messages on the device. Examples of messages captured on the device were parts of popular of funny songs, funny phrases or weird voices. It often occurred that people quickly pressed the recording button when they passed the device to record such trivial messages. This raised the awareness that lowering the threshold to record and play messages and utilizing the convenience of voice messaging might lead to a higher quantity of informal messages. The knowledge gained by the probe study supported several design decisions that have led to the concept created during this project.

Idea Development

Based on the outcome of the probe study, we elaborated on the basic ideas previously generated and utilized a variety of creativity techniques to create several versions of the concept. Based on the mid-level ideas shown in Fig. 4.5, finally one direction was selected and developed into a working prototype (Fig. 4.6).

Fig. 4.6 The fully functional final prototype had three messaging tokens and a simplified docking station

Family Circles

Family Circles, is a flexible and portable voice messaging solution that enables people to record messages and leave them at any desirable place in the home. Multiple, portable messaging tokens are able to store and play a single voice message and communicate secondary information utilising various properties of the light that is integrated in the tokens. Tokens containing a message that is already properly received, can be collected and stored together at a docking station. This dock also facilitates the recording of the audio messages onto the tokens and charging the tokens' batteries (Fig. 4.6).

The light in the tokens can vary in colour and brightness. The colour of the light can be altered at the docking station, in order to communicate meta-information about the sender, intended receiver or nature of a message for example. The brightness of the light can be changed repeatedly at the token itself, for instance to show one's appreciation for a message by making it more salient to others.

When someone encounters a messaging token, the stored message can be played by pushing down on the top of the token. This is a quick and intuitive way to play a message that maintains a low threshold for playing a message. This is crucial for other family members to become involved in the (on-going) indirect communication in the home.

Design Rationale

Using Light to Convey Meta Information

Coloured led lights with varying brightness can be used to convey different types of information. The colour could for instance indicate the intended receiver, while the brightness could easily be used to indicate the urgency of a message or how long ago it was created (as was the case in the technology probe system). One could even use it to indicate how much a message is appreciated by others, creating positive feedback to encourage leaving messages (a direct analogue to social media sites that let users indicate appreciation with just a click). Light signals can provide users with limited but useful meta-information about a message. For example they can identify the sender or intended receiver of a message, the nature or urgency of a message, its recency, or the appreciation for a message can be visualised by different colours or brightness levels of light. E.g., by making an appreciated message brighter, it will become more salient to others but also indicate to the creator of it, that it is appreciated and thus stimulating him or her to do it again. After some initial explorations we opted for using colour in an open ended way letting users set and change the colour of the light at will without any set semantics. This way, the family members have the possibility to attach their own meaning to the colour and design their own and possibly evolving conventions around it. Similar to the findings in Judge, Neustaedter, and Harrison's chapter in this book, families will get accustomed to each other's habits and patterns to form their own system around it.

Swarm Size

It is hard to comment on the number of tokens that is required for the smooth usage of the concept. This is dependent on the size of the family and the frequency of the indirect communication within a household. It is likely that a longitudinal study would provide clarity about this but as this project did not offer the opportunity to do so, no conclusions can yet be drawn regarding the required number of tokens. What can be said is that there should be multiple tokens so that messages can exist alongside each other and empty tokens can be used to record a reply to an existing message.

Appearance and Interaction

The main feature of the portable messaging tokens is to record a message and to be able to play it. To this end, the step to actually play a message should present a very low threshold and the token's physical shape should afford this need. Several shapes

Fig. 4.7 A visualisation of the messaging token. The left token shows the light ring, that is covered in the right one when the token is pressed to play a message

were explored and eventually we opted for a messaging token without push buttons, giving the token a clean look and a simple, intuitive operation.

The resulting shape (Fig. 4.7) is almost one big play-button, somewhat referring to the well-known emergency button that is both intuitive and easy to operate. In order to play a message, one presses the top of the token. By twisting the token's upper and lower body one is able to control the brightness of the light. Key to the success of the concept is the low threshold of playing a message. To this end, the functionality and interaction design have been deliberately kept clear and simple.

Creating Message-Threads at the Docking Station

Creating a visual overview of messages and their relation to each other can encourage people to join in a conversation. Creating a docking station for the messaging tokens could facilitate the collection of messaging tokens at a central place that, as was learned from the user study, is already a part of current indirect messaging habits. But apart from offering a place to collect the tokens, the docking station serves also for charging docked tokens which is an additional motivation for users to collect tokens at the docking station.

Evaluation

A brief field study was planned with the aim to evaluate the how the concept addresses the design goals stated earlier and whether and how it can become integrated into a family's messaging routines.

The field study involved two families fitting the target user group for the project. One family of four, with two children of 15 and 12 years of age and the other family of three, with one child of 18 years old.

Fig. 4.8 Field testing the functional prototype. This family positioned the dock at a well visible place in the living room

The families were invited to use sticky notes for 1 week and the Family Circles in the other week, allowing us to draw qualitative comparisons. Daily telephone interviews helped us keep track of their evolving experience and the usage patterns regarding Family Circles. Next to the docking station of the Family Circles, or the sticky notes collected in the reference period, a collection of postcards was given to participants too. These cards provided participants with the possibility to write down strong and/or weak points of the messaging solution that sprang to mind when using it.

The interviews revealed that the device was used quite regularly for varying purposes. Especially the father of the first family was keen on leaving informal messages to the rest of the family, often to no one in particular. Important to note here is that he did mention on the phone that he usually also leaves written messages. The mother indicated that she used the concept more to communicate messages of an organisational nature. The daughter did not leave any messages but also indicated she has never really left any messages in the home. She did say she liked to have her own colour and indicated the lights immediately made her aware of an awaiting message. Messages were left mostly around the docking station, at the dinner table or in the kitchen.

One family used colour to indicate for whom a message was intended while the other used it to indicate the creator of it. From the postcards and the debriefing interview was learned that this light was experienced as being useful as this makes a message very apparent and immediately shows who is supposed to hear it (Fig. 4.8).

Both families commented however that recording a message using the prototype as it was deployed then is somewhat of a hassle, which attributed to the poor usability of the docking mechanism. None of them actually changed the brightness of the light, indicating that this is not a crucial functionality. Due to the limited number of tokens we managed to produce and equip them with, they would overwrite messages on tokens often. This made the lifetime of a message too short for it to provide the intended benefits. Compared to how written notes were experienced, as a form of in-home indirect messaging, we could conclude that Family Circles were especially valued for their salience and their convenience of making messages more elaborate.

While we cannot draw generalizable conclusions from such a limited field study, this trial suggests that using Family Circles can support expressive, informal communication and that storing and distributing spoken messages is an appealing notion for intra family communication.

Conclusion

This chapter discussed how existing social media do not support expressive intrafamily communication. Our line of argumentation led us to identify the potential of indirect messaging as a way to connect individuals living in the same household but having very divergent daily routines. We presented the design and evaluation of Family Circles, a system that supports distributed voice messaging in the home, emphasizing the interlocked problems of designing the communication patterns, the interaction, and the form of the device.

Coloured light was used as an attractor for users to become engaged in the ongoing communication. Because of this and by creating a low threshold for listening to and creating a message, the system does appear to have the potential to increase the quantity of indirect communication; showing that this is the case requires a more extensive field study.

The system was open ended with regards to the semantics attached to lights, allowing families to assign idiosyncratic meanings to different light colours. Utilizing the emotional expressiveness of one's voice, voice messaging can be experienced as more personal than a written note. By lowering the threshold of responding to a message, the concept attempts to stimulate more informal messages and avoid oneway communication as much as possible.

The evaluation executed during this project could not, because of the limited time frame, offer strong evidence on the acceptance of the concept. Furthermore, because of the small number of tokens used in this evaluation, nothing conclusive was learned about the possibility to respond to an existing message. To this end, a more extensive concept evaluation is needed, that would be executed over a larger period of time and supported by a larger number of tokens. Furthermore, in order to get a realistic view on the quality of the concept, the connection between the token and the docking station should be improved in that recording a message will have a much lower threshold.

Family Circles, like Fida, Photomirror and Jakob, that preceded it seek not to just enable communication between family members, but to invite it, trigger it, trigger reactions to it, and even set the tone for the type of communication that will emerge. More than their functional characteristics these appliances do so by their form and the aesthetics of the interaction they support, pertaining to the nature of the physical actions that they require from their users, their fit to the space where they are used and their reciprocal influence upon the social context on which they are used. The designs share an open-endedness that allows users to appropriate them and use them in their own way, but also a pronounced simplicity regarding the functionality and the type of messaging that they support, filling a niche in a domain where rich media and always on connectivity are increasingly prevalent.

References

Breedveld, K., & Broek, A., Van Den (18 Oct 2006). Contacten met huisgenoten. Tijdsbesteding. nl: http://www.tijdsbesteding.nl/hoelangvaak/vrijetijd/contacten/huisgenoten/20061018.html. Retrieved 05 Dec 2010.

Breedveld, K., Broek, A., Van Den, Haan, J., de, Hart, J., de, Huysmans, F., & Niggebrugge, D. (2001). *Trends in de Tijd; Een schets van recente ontwikkelingen in tijdsbesteding en tijdsordening.* The Hague: Sociaal Cultureel Planbureau.

Brown, B. A. T., Taylor, A. S., Izadi, S., Sellen, A., Kaye, J., & Eardley, R. (2007). Locating family values: a field trial of the whereabouts clock. In J. Krumm, G. D. Abowd, A. Seneviratne, & T. Strang (Eds.), *Ubicomp, volume 4717 of Lecture Notes in Computer Science* (pp. 354–371). New York: Springer.

Dadlani, P., Sinitsyn, A., Fontijn, W., & Markopoulos, P. (2010). Aurama: caregiver awareness for living independently with an augmented picture frame display. *AI & Society, 25*(2), 233–245.

Davis, H., Ashkanasy, S., Benda, P., Gibbs, M., & Vetere, F. (2007). 'Time' and the design of familial social connectivity systems. *SIMTech: Proceedings of the International Workshop on Social Interaction and Mundane Technologies.* Melbourne, Australia, Nov 6–7.

Elliot, K., Neustaedter, C., & Greenberg, S. (2007). Stickyspots: using location to embed technology in the social pratices of the home. *Proceedings of the 1st international conference on Tangible and embedded interaction (TEI '07).* New York: ACM.

Gaver, B., Dunne, T., & Pacenti, E. (1999). Design: cultural probes. *Interactions, 6*(1), 21–29.

Haan, J., de, & Broek, A. Van Den (2000). (Vrije)Tijdsbesteding. In K. Wittebrood & S. Keuzekamp (Eds.), *Rapportage Jeugd 2000* (pp. 25–46). The Hague: Sociaal Cultureel Planbureau.

Hemmings, T., Crabtree, A., Rodden, T., Clarke, K., & Rouncefield, M. (2002). Probing the probes. *Proceedings of the 2002 Participatory Design Conference* (pp. 42–50). Malmö: Computer Professionals Social Responsibility.

Hutchinson, H., Mackay, W., Westerlund, B., Bederson, B., Druin, A., Plaisant, C., Beaudouin-Lafon, M., Conversy, S., Evans, H., Hansen, H., Roussel, N., Eiderbäck, B., Lindquist, S., & Sundblad, Y. (2003). Technology probes: inspiring design for and with families. *Proceedings of the CHI 2003.* Fort Lauderdale: ACM.

IJsselsteijn, W. A., Baren, J., van, & Lanen, F. van (2003). Staying in touch: social presence and connectedness through synchronous and asynchronoys communication media. In C. Stephanidis & J. Jacko (Eds.), *Human-computer interaction: theory and practice (Part III). Proceedings of HCI international* (Vol. 2, pp. 924–928).

Judge, T. K., Neustaedter, C., & Kurtz, A. F. (2010). The family window: the design and evaluation of a domestic media space. *Proceedings of the 28th international conference on Human factors in computing systems (CHI '10)* (pp. 2361–2370). New York: ACM.

Kassenaar, P. (2009). *JAKOB.* Eindhoven: Eindhoven University of Technology.

Khan, V. J., & Markopoulos, P. (2009). Busy families' awareness needs. *International Journal of Human-Computer Studies, 67*(2), 139–153. ISSN 1071-5819.

Khan, V. J., Markopoulos, P., Eggen, B., & Metaxas, G. (2010). Evaluation of a pervasive awareness system designed for busy parents. *Pervasive and Mobile Computing, 6*(5), 537–558. ISSN 1574-1192.

Lenhart, A., Madden, M., Rankin Macgill, A., & Smith, A. (2007). *Teens and social media. The use of social media gains a greater foothold in teen life as they embrace the conversational nature of interactive online media.* Washington, DC: Pew Internet & American Life Project.

Markopoulos, P., Romero, N., van Baren, J., IJsselsteijn, W., de Ruyter, B., & Farshchian, B. (2004) Keeping in touch with the family: Home and away with the ASTRA awareness system. In CHI '04 extended abstracts on Human factors in computing systems (CHI EA '04). 1351–1354. doi:10.1145/985921.986062.

Markopoulos, P., Bongers, B., Alphen, E., van, Dekker, J., Dijk, W., van, Messemaker, S., Poppel, J., van, Vlist, B., van der, Volman, D., & Wanrooij, G., van (2005). The PhotoMirror appliance: affective awareness in the hallway. *Personal and Ubiquitous Computing, 10*(2–3), 128–135.

Metaxas, G., Metin, B., Schneider, J., Markopoulos, P., & De Ruyter, B. (2007). Daily activities diarist: supporting aging in place with semantically enriched narratives. *INTERACT 2007, LNCS* (Vol. 4663, pp. 390–403). Heidelberg: Springer

Mynatt, E. D., Rowan, J., Craighill, S., & Jacobs, A. (2001). Digital family portraits: supporting peace of mind for extended family members. *Proceedings of the SIGCHI conference on Human factors in computing systems* (pp. 333–340).Seattle: ACM.

Noller, P., & Bagi, S. (1985). Parent-adolescent communication. *Journal of Adolescence, 8*(2), 125–144.

Richardson, R. A. (2004). Early adolescence talking points: questions that middle school students want to ask their parents. *Family Relations, 53*(1), 87–94.

Sellen, A., Harper, R., Eardley, R., Izadi, S., Regan, T., Taylor, A. S., et al. (2006). HomeNote: supporting situated messaging in the home. *Proceedings of the 2006 20th anniversary conference on Computer supported cooperative work (CSCW '06)* (pp. 383–392). New York: ACM.

Taylor, A. S., & Swan, L. (2005). Artful systems in the home. *Proceedings of the SIGCHI conference on Human factors in computing systems (CHI '05)* (pp. 641–650). New York: ACM.

Yalvac, M., & Helmes, J. (2007). Fida. Technische Universiteit Eindhoven: http://w3.id.tue.nl/nl/studeren/masters_program/classes/fida/. Retrieved 10 Nov 2010.

Yarosh, S., & Abowd, G. (2011). Mediated parent-child contact in work-separated families. *Proceedings of the 29th international conference on Human factors in computing systems (CHI '11)*. New York: ACM.

Zoontjes, R. (2007). Ralph Zoontjes—interactive product designer.Fida: http://www.nothings.nl/fida.html. Retrieved 7 Dec 2010.

Chapter 5
Enriching Virtual Visitation in Divorced Families

Svetlana Yarosh and Gregory D. Abowd

Abstract Divorce is a traumatic disruption in the lives of families that puts both parents and children at risk for long-term emotional and social consequences. However, if the non-residential parent maintains a quality relationship with the child, many of these negative consequences are mitigated. Divorced families face substantial challenges in parenting while living apart, especially as geographic separation often makes in-person visitation more difficult. Many families are turning to *virtual visitation*—supplementing in-person visits with use of communication technologies such as videoconferencing. However, current communication technologies are often inadequate to support long-distance parenting. We discuss the needs of divorced families and how these may be addressed through design. We present a case study of a single intervention, called the ShareTable, aimed at enriching virtual visitation between parents and children who live apart. Finally, we discuss the challenges and opportunities of designing for divorced families.

Introduction and Motivation

It is becoming common for children to live apart from one of their parents. The 2008 U.S. Census found that 26 % of children live with just their mother or just their father, with marital separation being the primary reason (U.S. Census 2008). A synthesis of psychology and sociology literature on divorced families shows that both the parents and children in separated families tend to score lower on multiple measures of wellbeing and adjustment (Amato 2001). However, the findings also

S. Yarosh (✉) · G. D. Abowd
Georgia Institute of Technology,
School of Interactive Computing, GVU Center,
Atlanta, GA, USA
e-mail: lana@cc.gatech.edu

G. D. Abowd
e-mail: abowd@gatech.edu

C. Neustaedter et al. (eds.), *Connecting Families,*
DOI 10.1007/978-1-4471-4192-1_5, © Springer-Verlag London 2013

suggest that when the remote parent and child maintain meaningful contact many of the negative consequences of separation are mitigated (Amato 2000). Unfortunately, contact with the remote parent drops precipitously after the first year of separation, often due to geographic separation (Seltzer and Bianchi 1988). Considering how difficult meaningful parent-child communication may be even in a collocated setting (as described in the chapter on intra-family messaging with Family Circles), it is not surprising that currently available remote communication technologies are often not sufficient to achieve the quality and quantity of contact necessary for long-distance parenting (Yarosh et al. 2009a). The challenges of remote contact are additionally compounded in the younger age group (6–13) targeted by the investigations described in this chapter.

Increasingly, families are seeking out alternative forms of synchronous and asynchronous communication to provide contact between visits. Successful attempts at leveraging tools like videoconferencing and instant messaging for remote parenting have drawn attention from the news media. The *New York Times* had several recent articles about videoconferencing with children (Conlin 2009; Harmon 2008). A number of recent publications have featured articles on *virtual visitation*—using communication technologies to augment face-to-face time between parents and children in divorced families (e.g., Flango 2003). There are efforts to incorporate virtual visitation into family law in almost every state, with five states already having added provisions for virtual visitation to custody case law (Cron 2006). Remote parenting is a relevant issue to families, lawmakers, and technology designers and is ripe for investigation from an HCI perspective.

In this chapter, we begin with a discussion of the specific challenges and unique aspects of designing for parenting after divorce. Next, we outline some opportunities for technological interventions in this space. We follow with a discussion of a case study of an intervention. We describe the ShareTable system, which aims to enrich virtual visitation in divorced families and summarize a formative evaluation of the system. Finally, we close with a discussion of the opportunities and challenges of doing work in this space.

Designing for Parenting After Divorce

In this section, we highlight the unique aspects of designing for parent-child relationships in divorced families. First, we discuss how designing for this relationship is different from designing for other close ties. Next, we talk about the specific challenges faced by divorced families in maintaining parent-child contact. Finally, we highlight opportunities in leveraging technology to support these families.

Designing for Parent-Child Relationships

Designing for parents and young children requires a different approach than doing so for friends or adult family members due to the (1) asymmetry in goals and needs between the parent and child, (2) the challenges posed by the cognitive and emotional limitations of young children, and (3) the focus on play and care rather than direct communication.

While strong-tie relationships (e.g., marriage) often involve symmetric goals and an equal involvement in relationship maintenance (Vetere et al. 2005), the parent/child relationship is characterized by asymmetry. Dalsgaard et al. (2006) found that the parent carried a greater responsibility over maintaining the relationship by creating a setting for trust and unity, providing care, and participating in play. Children rarely verbally expressed affection and they self-disclosed less than their parents desired. Modlitba and Schmandt (2008) and Yarosh and Abowd (2011) conducted interviews with work-separated families to find that parents and children have different emotional responses to separation; children are likely to experience anxiety before the parent leaves, whereas the parent is more likely to experience a sense of guilt during the absence. We conducted semi-structured interviews with parents and children in divorced families to understand the challenges that they faced in maintaining closeness (Yarosh et al. 2009a). Sharing on the part of children was oriented toward the current moment; if they were unable to share something when it occurred, they were unlikely to remember to do so in the future. On the other hand, parents were more concerned about interrupting the routines of the other household and were unlikely to contact the child spontaneously. All of these points highlight that parents and children have different approaches to their mutual relationship. Technology for these relationships must balance the needs and motivations of disparate participants to succeed.

Designing for children holds another challenge: the child's cognitive and emotional limitations may make long-distance contact difficult. As the child develops, he or she can begin to separate mentally from the here and now to imagine past and future events, comprehend how others see the world, and understand representational images of the world. Modlitba and Schmandt (2008) found in their interviews that it might be difficult for a young child to visualize where their parent is traveling and how long he or she will be away. Preschool children in interviewed families required the assistance of a collocated caregiver to initiate and make sense of their interaction with the remote parent. Even with school-age children, long-distance contact is challenging because many of them have not yet developed the communicational competencies to participate meaningfully in conversations without shared visual context (Stafford 2004). Lastly, children have limited attention resources and motivation for remote contact, so families often find it difficult to keep a remote communication session engaging enough to hold the child's attention (Ballagas et al. 2009).

Lastly, one of the distinctive characteristics of the parent/child relationship is that closeness is built more through play and care together than through conversation. As Ballagas, Kaye, and Raffle discuss in the chapter on remote reading with children, shared activities are a key characteristic of parent-child contact. Perhaps this is

unsurprising, since children have been shown to spend less than a 1 h/week participating in "household conversation" but more than 20 h/week participating in playing,
reading, studying, and hobbies (Hofferth and Sandbeg 2004). Dalsgaard et al. (2006)
found that parents and children build intimacy through care and play. Children and
parents participate equally in mutual play, in collaborative activities (doing a puzzle,
reading, or cooking together), in playing with shared artifacts (action figures or a
board game), and in physical play behaviors. On the other hand, care is unidirectional
from the parent to the child and includes activities such as setting rules, providing
resources for learning, giving physical care, and assisting with everyday tasks and activities. Development literature emphasizes the importance of parental involvement
in both care and play activities, to build secure relationships (Kelly and Lamb 2000).

Divorced Family Dynamics

In all parent-child relationships, continued quality and quantity of contact is key to
building a connection but is rarely achieved in divorced families. We describe the
challenges faced by these families.

Separation carries significant negative consequences for both the child and the
parents (Amato 2001). However, these negative consequences can often be mitigated if the distributed parent stays instrumentally involved in the child's life (Amato
2000). Smyth (2002) emphasizes that the quality of contact may be as important to
explore as the quantity. "Quality contact" may be difficult to unpack, but developmental psychologists have used the term "authoritative parenting" to describe the
combination of monitoring and support that is likely to lead to positive behavioral
and academic outcomes for children (Smyth 2002). Gray and Steinberg (1999) isolated and examined the behaviors that characterize this construct to find that the
amount of communication and the act of showing interest in the child's life were the
most influential constituent behaviors involved in authoritative parenting. Additionally, frequency and variety of contact are also important to maintaining relationship
quality. Kelly and Lamb advise that parenting arrangements should provide "opportunities to interact with both parents every day or every other day in a variety
of functional contexts" (Kelly and Lamb 2000). Unfortunately, these prerequisites
for quality contact may be difficult to achieve for parents and children who live
apart. Furstenberg and Nord (1985) studied patterns of parenting after separation to
show that the distributed parent was likely to be involved socially in the child's life,
but rarely set rules or assisted with care activities such as helping with homework.
Seltzer and Bianchi (1988) showed that the quality and quantity of contact with the
distributed parent decreased dramatically after the first year of separation.

We conducted an in-depth interview study with 15 residential parents, non-
residential parents, and children from divorced families to better understand the
practical challenges they face in everyday life (for a more complete presentation
of these results see Yarosh et al. 2009a). The two major struggles experienced by
these families center around maintaining a shared context while living apart and
managing conflict. First, the remote parent often faces challenges in staying aware

of the child's state and activities. Children are often not very good in providing such information and the residential parent may not be motivated to keep the non-residential parent up-to-date. Second, parents often have to weigh the desire to contact the child with the possibility of interrupting the daily routines of the other household. This often leads to most communication being scheduled ahead of time. Finally, parents often struggle with seeding conversation and keeping the child engaged. On the other hand, children struggle with managing the competition over their time and affection between the parents. In our study, we found that children were much more aware of this competition than their parents anticipated. This uncomfortable situation is often exacerbated by a lack of a private space to communicate with the remote parent. Lastly, the fact that most remote interaction is scheduled makes it difficult for children to communicate spontaneously when they think of something they want to share. Often, by the time the time there is an opportunity for scheduled interaction, the thought or feeling is long forgotten.

The themes we identified (which were confirmed in other work) (Odom et al. 2010) suggest that members of divorced families balance two major goals: reducing tensions between households and maintaining closeness. Children may try to reduce tensions by keeping the details of their involvement with the other parent as private as possible. Parents may seek to reduce conflict by maintaining only minimal contact with each other, respecting each other's autonomy, and minimizing unscheduled interruptions of the other household. However, both of these goals may conflict with the parents' desire to remain aware of the child's everyday activities to provide support and drive conversation. The parent's need to minimize interruption may also clash with the child's goal of achieving spontaneous contact, as it leads to a regimented schedule of interaction with few opportunities for spur-of-the-moment conversation. Both parents and children expressed that they would prefer to stay in touch through something richer than phone conversations, but found that asymmetric rules and asymmetric access to infrastructure between households often lead to the lowest common technological denominator. While the non-residential parent may be driven to upgrade the infrastructure, there is often little motivation for the residential parent to do so. The residential parent may see the introduction of a new communication technology as a violation of their autonomy in raising the child. While all parties share the common goal of achieving positive outcomes for the child, they may disagree on what constitutes a "positive outcome" and how to get there. Designing for divorced families requires maintaining the balance between building closeness and reducing tension in such a way that the technology can be acceptable to all members of the family.

Current Use of Technology in Divorced Families

Though there are few studies investigating the effect of available communication technologies on maintaining contact between parents and children, the Pew report on the American "networked family" (Kennedy et al. 2008) showed that such technologies do have the potential to raise the quality of communication with friends and family. Fifty-three percent of respondents indicated that mobile phones and the Internet have increased their quality of communication with friends and distributed family (44 % said that it remained the same). The report also indicated that increases in time spent using social media comes at the expense of time spent watching television, not at the expense of time spent socializing in-person. Most families already have the infrastructure to use communication technologies such as videoconferencing and many seem to be excited by the opportunities provided by these media.

Non-residential parents often turn to technology to supplement in-person communication. Some parents maintain websites and forums dedicated to sharing ideas about using technology to stay in touch, such as distanceparent.org and internetvisitation.org. Particularly, the combination of telephone, videoconferencing, and instant messaging to supplement in-person visits is known as *virtual visitation* (Flango 2003). As of 2009, five states have passed laws allowing virtual visitation to be incorporated into custody decisions. Several family law periodicals have featured virtual visitation, stating, "technology may be able to help maintain a relationship that would otherwise cease" (Shefts 2002). Despite the fact that it is already becoming incorporated into state law, there has been relatively little academic or industry research into virtual visitation.

In our interview study (Yarosh et al. 2009a), we found that technology use in divorced families is often characterized by asymmetric access to infrastructures between the two households, which often leads to the lowest-common-denominator interaction. Unfortunately, this often means the telephone. Both the children and parents in our study found audio-only communication inherently difficult and unsatisfying (also confirmed in other investigations Ballagas et al. 2009). Most conversations amounted to quick calls good night or quick updates. While several families reported that videoconferencing was a much richer way of interacting, few used in regularly. Videoconferencing is difficult to set up (Ames et al. 2010), often requires more technical savvy and motivation than one or both parents in divorced families are willing to provide, and introduces concerns over privacy and safety that may prevent its adoption. Despite the widespread popularity of Skype, videoconferencing is still not used routinely for remote parent-child content. For example, in a study published in 2011, out of the 14 families where parents frequently travelled for work, only 9 had tried videoconferencing and of those only 5 used it regularly (Yarosh and Abowd 2011). Despite widespread availability of free services like Skype, videoconferencing still presents very really challenges for the majority of families.

Overall, it seems that divorced families are open and willing to consider new technologies but there are few technologies are designed explicitly for their needs.

Potential for Technological Intervention

There are many opportunities for design interventions to support divorced families. In this section, we provide an overview of opportunities clustered from our work (Yarosh et al. 2009a) and that of Odom et al. (2010).

In the previous sections, we have shown that care activities and instrumental parenting on the part of both residential and non-residential parents are important to the child's wellbeing. Unfortunately, there are currently limited opportunities for the non-residential parent to provide such care. For older children, providing remote homework help may present one opportunity for instrumental contact. There is a great deal of CSCW and HCI literature on supporting work remotely that can be leveraged for homework help. Additionally, consistent instrumental care can only be possible if parents who share joint custody maintain consistent rules and cultures across households. Odom et al. (2010) suggest that photo sharing, shared calendaring, and online networking can provide opportunities for creating a "joint culture" without direct communication between the parents.

Objects can hold a great meaning for children when their life is disrupted by divorce. An object brought between households (such as a teddy bear) can provide a necessary sense of stability. Other objects (such as a soccer ball or a favorite photo) can remind of shared time and reinforce closeness. Everyday physical objects could be augmented to support a sense of connection and closeness when direct contact between the parent and child is impossible. For example, the child's augmented soccer ball could vibrate slightly when her remote dad is playing soccer, encouraging her to participate in the same activity. Alternatively, virtual possessions could become a thread of stability by providing a context that is available to the child regardless of his or her physical location (Odom et al. 2010).

One of the biggest needs for divorced families is creating new opportunities for remote contact. One way to do this is by supporting asynchronous interaction. There are currently very few opportunities for remote communication with children, since they rarely own mobile phones. Creating dedicated messaging devices for children or incorporating such features into existing portable gaming devices would allow for quick spontaneous contact even when either party is unavailable for synchronous contact. The second way of creating new opportunities for remote contact lies in empowering the child to initiate the connection without help from the residential parent. The child is aware of the competition between the parents over his or her time and affection and may hesitate to approach one parent for help in setting up the connection to the other parent. Making it possible and safe for even young children to use technology like videoconferencing would increase opportunities for interaction. Lastly, we could focus on increasing the length of the synchronous communications between parents and children. In order to help parents and children have more meaningful interactions, it would be useful to provide the parent with information about the child's everyday life and activities to help seed the conversation. While in intact families, the remote parent can rely on a local adult to provide this information (Yarosh and Abowd 2011), divorced families may benefit from more indirect sources of information such as awareness systems. Finally, in order to make communication engaging and meaningful to both participants, it is helpful to provide a shared context for the interaction, especially when that context can include care or play activities.

While there are a number of possible interventions for divorced families, the remainder of this chapter focuses on a case study of one possible intervention. The

ShareTable is a technology to support richer and more engaging remote synchronous interaction between parents and children.

The Road to the ShareTable

We focus on designing a technology to support richer synchronous interaction between parents and children in divorced families. In the next sections, we describe the specific design requirements that drove the creation of the ShareTable system, provide a brief overview of the system implementation, report on an initial evaluation, and discuss the process of adapting the system for a long-term field deployment. A more detailed discussion of this work can be found in (Yarosh et al. 2009b).

Design Requirements

From our interviews with divorced families and the previous work in this domain, we determined four design requirement for a synchronous remote communication system for parents and children that face separation due to divorce.

1. **Add a Visual Channel for Communication**
   The most common theme reported by both parents and children in our interview study was dissatisfaction with audio-only communication. During the middle childhood, children are still developing the conversational competencies to interpret irony, humor, and fantasy (Stafford 2004). Providing multiple channels and modalities for communication, particularly video, affords additional cues for the child and provides a shared context for communication.
2. **Function without a collocated Adult's Help**
   The families we interviewed did not use videoconferencing regularly, because most videoconferencing systems are complex enough to require a collocated adult's involvement to arrange a chat session. Additionally, some parents saw it necessary to supervise videoconferencing, since the child could potentially contact or be contacted by a stranger. Our goal is designing a dedicated communication system with a minimal control interface that reduces the need for a Collocated adult to assist the child with setting up and maintaining the connection.
3. **Support a Wide Variety of Play Activities**
   Keeping the child engaged and seeding conversation were two major challenges reported by parents. We seek to support engagement by leveraging activities that the parent and child are already used to doing together. We emphasize the system's ability to support a variety of activities, rather than incorporating interfaces for specific games or requiring specific accessories.
4. **Provide Opportunities for Care Activities**
   There is strong evidence that instrumental involvement of both parents in raising the child correlates with positive outcomes for children (Kelly and Lamb

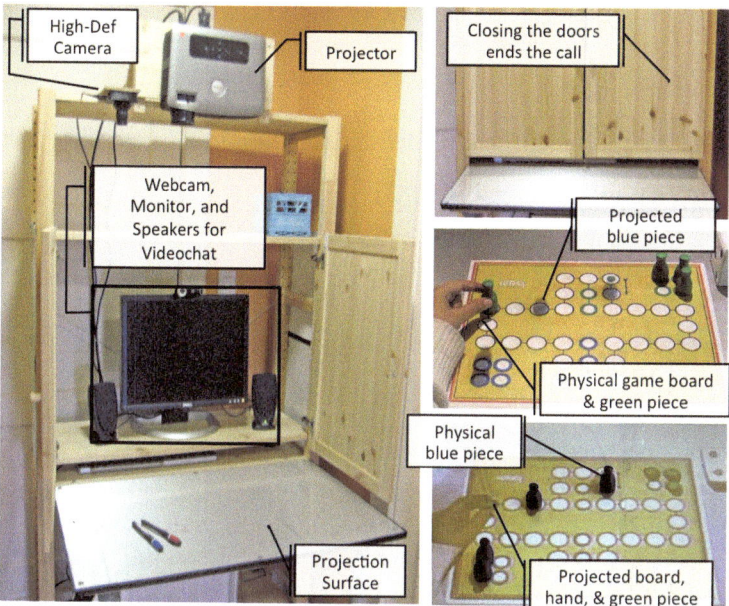

Fig. 5.1 The ShareTable system consists of standard videoconferencing and a shared tabletop created through top-down projection, which allows joint activities with physical artifacts, such as board games and worksheets

2000). Many care activities require physical presence; however, there is a clear opportunity for remote instrumental care in providing homework assistance. The challenge to us as designers is to afford transitions between the physical artifacts of homework that the child possesses (e.g., textbook, worksheet) and digital versions of these artifacts, which the parent can view and annotate. We discuss how we addressed this challenge in the next section.

In the next section, we describe the ShareTable system, which is meant to address these four design requirements.

System Overview

The ShareTable system consists of two identical table setups in the households of the child and the remote parent. Each shared workspace consists of an overhead camera that records any activity over the surface of the table and a projector that displays this video on the paired table in the other home (see Fig. 5.1). The video from each camera is aligned precisely with the projection, so that artifacts placed on one table appear projected in the same location on the other table. The tabletop is coupled with a videoconferencing system (i.e., monitor, webcam, speakers, and microphone) that let the users see and hear each other "face-to-face." As in other videoconferencing systems, each user also sees a smaller video window showing how they appear to the other person. This setup allow the parent and child to talk to each other while *doing* something together, such as helping with homework, playing with plastic action figures, drawing, etc. We took the approach of sharing direct video rather than creating specific content to be shared (in contrast to the reading together chapter) in order to support play and collaboration with *any* toys, books, or artifacts that the parent and child may already have around the home.

The basic idea behind the ShareTable is simple, but multiple implementation questions had to be addressed in developing a functioning prototype. First, we needed an alternative to most existing tabletop systems because we wanted to support layering physical artifacts. To solve this, we chose to implement the system using top-down projection. For example, if the parent places a physical token on a projected game board, top-down projection allows the projected token to appear on top of the child's physical board rather than projected unseen on the board's bottom. Similarly, if a parent writes a comment on top of a projected worksheet, top-down projection allows this annotation to be displayed on top of the physical worksheet. Second, we needed to solve the problem of visual feedback or "echo," which is a major concern in camera-projector systems. Unmodified, the camera records an image of the projected artifact and sends it back to the originating surface. If the physical artifact is moved, an echo of its projection remains on the surface. If projected images are re-captured without any intervention, the resulting image keeps getting brighter and less clear. Without some way to filter projected artifacts from real ones, the ShareTable would be unusable due to this feedback effect. We wanted a lightweight way to eliminate visual feedback that still preserved color, so we used linear polarizing lenses to filter out the projected artifacts from the physical ones. Light that passes through the lens becomes polarized and cannot be seen through a lens with the opposite polarity. Thus, by attaching lenses with perpendicular polarization to the camera and projector, we prevent artifacts from being re-projected. In order to preserve the polarization of the light once it strikes the table surface, we use a non-depolarizing silver lenticular projection screen as the surface backdrop.

Initial Evaluation

Our initial evaluation took place with an early prototype of the ShareTable that included both the face-to-face and the shared tabletop surface, but did not transmit data over the network (the tables were connected via video cables) and did not address how the connection would be initiated (we set up the connection for the participants). Though lab-based evaluations are inherently limited, the questions that motivated this initial investigation could reasonably be approached in a controlled setting. First, we wanted to observe ways in which interaction with the ShareTable is different from plain videoconferencing. Second, we wanted to establish that children would be able to understand and manage the interweaving of physical and projected spaces created by the ShareTable. Finally, we were interested in exposing participants to our system to gain insight to potential activities they may want the system to support. The lab evaluation methods and results are described in more detail in (Yarosh et al. 2009b).

Methods

Seven parent-child pairs participated in the study. The set of parents, four males and three females, varied in age from 30 to 44 (average 37.3, median 38). Their occupations ranged from attorney to professor to student, but all had a high degree of education. The children, three females and four males, were between 7 and 10 years old (average 8.4, median 9). We recruited these participants through word-of-mouth and flyers posted around campus, which advertised that we were looking for individuals interested in technology for families who spend significant time apart.

We familiarized participants with the residential lab where the study took place and introduced the project. We gave them time to play and experiment with the ShareTable in an unstructured manner. They were encouraged to think about how they would possibly use such a system while apart and to actively try out some of those activities. When the participants were ready to continue, we asked them to perform three separate tasks and fill out a brief questionnaire.

The first two tasks involved completing a worksheet together. The worksheet given to the child consisted of a political map of Africa without any labels, with instructions to color in all countries that began with a certain letter ("M" in the first task, "A" or "Z" in the second). The parent was given an answer sheet—a colored map of Africa that contained the names of the countries and their capitals—and instructed to assist the child in any manner they thought appropriate. For one of these tasks, the parent-child pair was asked to use videoconferencing, while the other task allowed them to use the ShareTable. Each parent-child pair completed both tasks, representing a within-participant design, counter-balanced for order effects. We were interested in comparing the strategies that parents and children used with the addition of the extra video channel.

In the third task, the parent and the child were asked to play a board game together using the ShareTable system. This represents a task that is currently impossible to carry out using a videoconference system alone, so there was no videoconferencing condition. We provided a simple game, based on the idea of "Ludo" or "Sorry!" (see Fig. 5.1). Only the parent's side had the physical game board, but each side had physical token pieces and a die. Thus, the child had to place his or her pieces on the projected surface of the board. We were interested in whether the child would be able to manage turn taking and access in this unusual space, which interweaves physical and projected artifacts. We chose the game of Ludo because it includes a rule that if your opponent stops her token in a space that's currently occupied by one of your tokens, you must move your token back to the start position. In a physical game, this rule is easily enforced with physical constraints (only one piece can occupy a give space), but we were interested in seeing how the game would play out over the ShareTable where no such constraints were present.

We asked each parent-child pair to commit 1 h to this study; however, they were also given the option of continuing to play with the system in whatever way they chose at the end of that time.

Comparing ShareTable and Videoconferencing

We began by observing how parents and children completed two worksheet tasks—one with the ShareTable system and the other with plain videoconferencing. After completing each task, we asked them to answer a few questions about their experience. A more detailed description of these findings can be found in Yarosh et al. (2009b).

We asked each parent and child how difficult it was to do the worksheet with each communication medium and how much he or she liked using each system on a 5-point Likert scale. We hypothesized that the ShareTable would be rated as both easier and better liked than plain videoconferencing and this was supported by the data. We turned to the observation data to better qualifying this difference.

In the videoconference condition, children and parents used the following strategy: the parent would verbally explain where the country is (e.g., "the little one to the left of the big one that looks like a heart"), the child would point to the country and hold up the worksheet to the webcam, the parent would confirm or reject the selection, and the child would color in the country if it was confirmed. The main breakdown in the process occurred as the child tried to identify and confirm the country. Two of the children seemed to assume that the parent could see where they were pointing without holding up the paper (even though it was explained that the parent could not). Five of the children had trouble understanding how the worksheet would look to the parent when held up to the camera—holding it too close, too far, or even upside down.

In the ShareTable condition, the child would keep the worksheet flat on the table. The parents described the correct country verbally, by pointing to it with their finger, or by circling it with a marker. Children would verbally confirm if they had the right country or would touch the country with the tip of the marker and look up at the video screen for confirmation. Interestingly, parents did not seem to be concerned with the efficiency of completing the worksheet. None of the parents simply put the sheet with the answers on the table. In one family, the mother explicitly acknowledged that if she showed the answers, she would feel like she was cheating and that her son would probably learn more if they worked through the worksheet together. Another common behavior was taking verbal tangents from the task to tie the worksheet to other experiences in the child's life. For example, a father pointed to an African country to tell the daughter a story about her aunt who currently lives there. Additionally, every parent made a remark about the country Madagascar and the children's animated movie by the same name.

It has previously been demonstrated that gestures over video streams can support quicker completion of remote tasks. When one user assists another for work, measures like time to completion make a lot of sense. However, when the users are parents and children, completing the task takes a back seat to engaging with each other. In the ShareTable condition, we noticed a greater level of engagement between the parent and the child. They spent more time looking at each other and less time looking at the task. They also spent more time laughing and talking about peripherally related information. Parents supported their child's learning not by making sure that the worksheet was completed quickly, but rather by tying the activity to other aspects of the child's life, such as familiar children's media. By making the logistics of the task easier, we conjecture that the ShareTable freed the parent and child to focus on these other aspects of communication. In other words, the ShareTable enriched the activity of remote homework help.

Using the ShareTable to Play a Physical Board Game

To see how parents and children coordinated turn taking and interaction with physical artifacts while using the ShareTable, we asked them to participate in a simple board game task, similar to "Ludo." Since the ShareTable just projects a video stream, each participant can only physically manipulate the artifacts on his or her side of the table. We wanted to see how participants would manage the interaction of "bumping" each other's pieces back to start. While all but one parent-child pair explicitly verbally acknowledged the possibility of refusing to move their piece when bumped, but quickly dismissed it, as it would "ruin the game" or make the game "no fun." In fact, there was a great deal of physical behavior surrounding the bumping of a piece despite the fact that the participants could not physically replace the opponent's piece back to the start. A common behavior was manipulating the game token in a "dancing" motion on top of the projection of the opponents' piece after bumping an opponent.

Unlike an online board game, the ShareTable leaves the management of turns and rules up to the users. While the user was taking his or her turn, they would usually focus on the table surface; however, during their opponents turn, they focused on the face-to-face video. Looking up at the screen at the end of one's turn seemed to signal to the other person that it was his or her move. One interesting facet we observed was that parents tried to bend the rules of the game to the advantage of the child—children won six out of the seven games played. Parents would do this by giving the child strategy advice and by letting them re-do moves or take extra turns. If we had built explicit games and rules into the infrastructure of the ShareTable, this interaction may have been lost.

In post-task interviews, two of the parents explicitly mentioned that, despite the lack of access to the opponent's pieces, playing the board game using the ShareTable felt much more similar to playing a board game in-person than using any other computer-mediated channel. Another parent mentioned that after the first 10 min of using the ShareTable, he felt that he could focus entirely on interacting with his daughter, rather than "using the system." All of the children we interviewed said that they would like to try more board games with the ShareTable. Two of them explicitly requested the chance to play again at a later time. To summarize, parents and children were successful at managing access to artifacts and turn taking without specific system support—they mutually acknowledged the rules and possibilities of the interface and acted to manage them in a way similar to in-person interaction. Playing a board game using the ShareTable was more similar to the rich experience of playing together in-person than to the controlled experience of playing an online game.

Observing Free Tasks and Considering Future Possibilities

Finally, we observed the way users interacted with the ShareTable when given an opportunity for free play before and after the tasks. We sought to identify the features of the ShareTable that supported or hindered the activities that the parents and children chose. Several parent-child pairs participated in "collaborative drawing" in which the child or the parent would initiate a drawing while the other added elements to it (e.g., child draws butterfly and the parent adds patterns on the butterfly's wings). One of the parents mentioned that this task was actually easier with the ShareTable than in-person because she and her son could occupy the central physical location at the table without getting in each other's way. We observed a variety of other playful activities. One parent-child pair participated in a "tracing" activity—the father put his hand on the table and the child carefully traced it. In one family, the child played a game of "tag" by trying to catch the projected version of her dad's hand with her own. One family really wanted to try doing their own task—playing a game of chess with their own board and pieces. They were successful, but we noted that because the ShareTable places the two users on the same side of the table, the father was put in the awkward position of having to play his pieces from the opponent's side of the board.

   In post-study interviews, we asked the parents and children how they would use the system in their own home and if they had any suggestions for modifying the ShareTable. One parent said she wanted her son to be able to leave a short note on the table when he gets home from school. She wanted to be able to access a message left on the table from her mobile phone to quickly get feedback that her son safely arrived at home. One child suggested that her father could put printed pictures on his side of the table so that she could trace them. Another child mentioned that he would have liked to be able to share the drawings he and his mother created by giving them to his father to take to work or hanging them on the refrigerator. Both parents and children said that they would use the ShareTable for both play and homework if they had one in their home. Several parents mentioned wanting to be able to read with the child, but three expressed a concern that the resolution of the ShareTable surface would not be high enough to allow comfortably reading most books. The ShareTable only provides the medium for the interaction—creating content is left up to each family—so, it was encouraging to see that our participants could come up with a variety of compelling use cases for the system.

From Functional Prototype to Robust System

While the lab-based evaluation demonstrated that the ShareTable was compelling for parents and children, there were a number of changes necessary in order to make the system ready for long-term deployment in the home. We present these here to demonstrate that the transitioning from "functioning prototype" to "robust system" is frequently not trivial.

The first step was converting our quick Python solutions into something that could stand up to everyday use by a real family. For us, that meant changing large parts of the system to leverage existing APIs. After some experimentation, we decided to use the Skype API for the face-to-face video and audio, while the tabletop video used the Axis Camera API. While we do gain robustness by working with existing APIs, there were several points at which the APIs did not support specific functions we needed, requiring creative workarounds.

At this point, we had a number of tradeoffs to make in the design of the system. While the lab-based prototype of the ShareTable avoided network issues by physically connecting the two tables, we needed to consider how this system would function over a real network. Even leveraging the efficiencies of existing solutions, we are attempting to transfer considerably more data than a household connection is capable of supporting. We found that $1,280 \times 1,024$ overhead camera image was the minimum to allow size 14 fonts to be readable over the table. When this is added to the already-heavy requirements of a Skype video call, most home networks come up short. In order for the system to work, we needed to consider potential trade offs to conserve bandwidth. Other videoconferencing systems do this by prioritizing frame rate over resolution. This makes sense for face-to-face videoconferencing where being able to perceive gesture and expression is paramount. Our face-to-face video adopts this strategy as well. However, for our tabletop surface we chose a different approach. In order to support reading and helping with homework, we decided to prioritize resolution over frame rate. With a home bandwidth connection, this unfortunately often means a frame rate as low as 2 fps.

Lastly, unlike the lab-based prototype, we needed to consider the way interaction would be initiated using the system. Most similar media spaces have been evaluated in the lab, therefore not needing to consider the way a connection would be initiated. Alternatively, many media spaces assume an always-on connection, again avoiding the question of initiating contact. An always-on connection would not be an acceptable solution to divorced families, so we needed to consider how to implement a solution for initiating a connection that would be simple enough for a child to use and where the state of the system would be immediately apparent to others in the house. We chose to implement a simple physical metaphor for initiating the connection. Opening the ShareTable cabinet activates the connection to the paired

Table (through a simple Reed switch circuit) and the receiving table rings as a telephone might. Closing the open doors ends the call.

While most telepresence studies are conducted in the lab, our process with the ShareTable emphasizes that there are a number of problems that are avoided in such deployments, but need to be considered in order to prepare a system for the field. The steps between "functioning prototype" and "robust system" are rarely made visible in publication, however we hope that by making these steps more transparent we can encourage others to try to take their system beyond the lab. We are now planning to conduct month-long deployments of the ShareTable system with three divorced families (6 households).

Discussion

In this section, we discuss the challenges and opportunities of designing for divorced families highlighting both the difficulties and the importance of working in this domain.

Challenges

The three most salient challenges of designing for divorced families are (1) creating technology before there is law to support its use, (2) designing in situations with conflicting stakeholders, and (3) taking technology from the lab into the home.

Though virtual visitation is in the process of becoming part of family law in most states, this is a slow process. While in the future a technology like the ShareTable may be installed at the request of the non-residential parent (for example, as a precondition for relocating the child), currently we can only deploy it in low-conflict families, where both parents are motivated to consent to this system. This doesn't allow us to fully explore the implications of the technology we have built. This may be the case for many novel technologies created for this audience, as the law will inevitably be slower than technological innovation.

Divorce is inherently a setting of conflict where different stakeholders may have radically different needs and motivations. Researchers in this domain acknowledge that divorce is an emotionally charged topic that is difficult to explore without "being identified as either a conservative or a liberal voice" (Amato 2000). Working closely with divorced families, there is implicit pressure from the participants to ally with a particular party. As an explicit design decision, we try to remain consistent with the shared goal of providing positive outcomes for the child. However, we must acknowledge that it is possible that introducing new technology in this domain may lead to unintended consequences and there are assumptions implicit in our intervention. We make the assumption that contact with both biological parents is beneficial to the child. While there is a large body of empirical evidence to support this hypothesis (e.g., Amato 2000; Wallerstein and Kelly 1996), this will not be true for every child and every parent. As with any divorce situation, it becomes the responsibility of policy makers, judges, and parents to tailor a solution appropriate to the specific situation. The most tentative assumption that we make is that improving communication between the child and the distributed parent will not negatively affect other family relationships in the child's life. There is evidence that quality contact with the biological parents does not negatively affect the child's relationship with their stepparents (Furstenberg and Nord 1985). However, it is difficult to predict the way new technologies will affect the lives of users. We seek to include nonuser stakeholders in the evaluation of new communication technologies to help us understand when such conflicts do occur. We hope that by being explicit about our assumptions and the values that we bring to the table as researchers, we can avoid the trap of false objectivity.

Lastly, designing for divorced families shares a challenge with all design for the domestic space. Technologies are difficult to take from initial prototype to working system and nothing less than a robust solution would support a reasonable evaluation in the home. With divorced families, it is perhaps more important to deploy in the home than in other domestic situations. Interventions for divorced families must become familiar and routine enough in the home that the families stop acting like "good participants" (Brown et al. 2011) and begin acting within the patterns that truly reveal the nuances of the family's interactions. Unfortunately, it is hard to make such long deployments work within the timelines and budgets of academic research.

Opportunities

Despite all of the challenges highlighted above, there is a lot to gain in designing for divorced families.

Studying divorce foregrounds family issues that are usually difficult to get at in other families. This allows us to study situations that may be more infrequent in other families and thus harder to see and consider in the design. The first of these issues is the one of conflict. While conflict is assumed in divorced families, intact families are often considered to be harmonious units with common goals and motivations. This is often not the case, and making this assumption can lead to communication breakdowns (Sillars et al. 2004). The second issue relates to non-consensus and asymmetrical motivation. The motivations of the child to communicate with his or her remote parent are likely be different from the expectations of both the residential and the non-residential parent. This highlights the importance of keeping in mind the obligation to communicate that new technologies may introduce and what may happen if these expectations are not met. The last issue concerns the privacy in families. While privacy may be a background issue in many intact families, we need to keep in mind that all families function within "numerous interrelated boundaries operating simultaneously" (Caughlin and Petronio 2004). As recent deployments of media spaces in the home have shown, families do not function as a single-minded unit regarding how they manage their privacy with other family members and violations of individual privacy preferences can lead to the rejection of a technological intervention (Judge et al. 2011).

Just as the medical field tends to focus on the most extremely affected patients as a case study, so too can divorce serve as a "worst case scenario" of family interaction. It is likely that technologies designed for divorced families can extend to other situations such as grandparent-grandchild interaction, work-separated families, or even incarcerated parents. Conversely, it is less likely that technologies designed for situations with minimum conflict will be able to flourish in high-conflict households. At the same time, divorce is currently the most common cause of parent-child separation and one of the most permanent ones. Addressing the needs of divorced families provides an incredible opportunity to create an impact in the lives of over a million children who experience divorce every year in the United States alone (Wallerstein and Kelly 1996).

Acknowledgements This work has received support from a number of sources including AT&T Fellowship, IBM Fellowship, Nokia University Award, and a kynamatrix grant. We gratefully acknowledge all of the people who have contributed their time and energy to the ShareTable project: Stephen Cuzzort, Brian Di Rito, Jee Yeon Hwang, Sanika Mokashi, Hendrik Müller, Hina Shah, Jasjit Singh, Shashank Raval, and Anthony Tang. Lastly, we want to thank all of our participants for their time and honesty.

References

Amato, P. R. (2000). The consequences of divorce for adults and children. *Journal of Marriage and the Family, 62*(4), 1269–1287.

Amato, P. R. (2001). Children of divorce in the 1990s: an update of the Amato and Keith (1991) meta-analysis. *Journal of Family Psychology, 15*(3), 355–370.

Ames, M. G., Go, J., Kaye, J., & Spasojevic, M. (2010). Making love in the network closet: the benefits and work of family videochat. *Proceedings of the CSCW* (pp. 145–154). New York: ACM.

Ballagas, R., Kaye, J. "J"., Ames, M., Go, J., & Raffle, H. (2009). Family communication: phone conversations with children. *Proceedings of the IDC* (pp. 321–324). New York: ACM.

Brown, B., Reeves, S., & Sherwood, S. (2011). Into the wild: challenges and opportunities for field trial methods. *Proceedings of the CHI* (pp. 1657–1666). New York: ACM.

Caughlin, J. P., & Petronio, S. (2004). Privacy in families. *Handbook of family communication* (pp. 379–412). Mahwah: Lawrence Erlbaum.

Conlin, J. (2009). Living apart for the paycheck. *The New York Times.*

Cron, S. K. (2006). Virtual Visits: New Law Provides Alternative Visitation Options. *Law Office Computing* (pp. 28–29).

Dalsgaard, T., Skov, M. B., Stougaard, M., & Thomassen, B. (2006). Mediated intimacy in families: understanding the relation between children and parents. *Proceedings of the IDC* (pp. 145–152). New York: ACM.

Flango, C. R.(2003). Virtual Visitation: Is This A New Option for Divorcing Parents? *The 2003 Edition of the Report on Trends in the State Courts* (pp. 20–22).

Furstenberg, F. F., & Nord, W. C. (1985). Parenting apart: patterns of childrearing after marital disruption. *Journal of Marriage and the Family, 47*(4), 893–904.

Gray, M. R., & Steinberg, L. (1999). Unpacking authoritative parenting: reassessing a multidimensional construct. *Marriage and the Family, 61*(3), 574–587.

Harmon, A. (2008). Grandma's on the computer screen. *The New York Times.*

Hofferth, S. L., & Sandberg, J. F. (2004). How American children spend their time. *Marriage and Family, 63*(2) (121AD), 295–308.

Judge, T. K., Neustaedter, C., Harrison, S., & Blose, A. (2011). Family portals: connecting families through a multifamily media space. *Proceedings of the CHI* (pp. 1205–1214). New York: ACM.

Kelly, J. B., & Lamb, M. E. (2000). Using child development research to make appropriate custody and access decisions for young children. *Family Court Review, 38*(3), 297–311.

Kennedy, T. L. M., Smith, A., Wells, A. T., & Wellman, B. (2008). Networked families. *Pew Internet & American Life Project Report.* Washington, DC.

Modlitba, P. L., & Schmandt, C. (2008). Globetoddler: designing for remote interaction between preschoolers and their traveling parents. *Extended Abstracts of CHI* (pp. 3057–3062). New York: ACM.

Odom, W., Zimmerman, J., & Forlizzi, J. (2010). Designing for dynamic family structures: divorced families and interactive systems. *Proceedings of the DIS* (pp. 151–160). Aarhus: ACM.

Seltzer, J. A., & Bianchi, S. M. (1988). Children's contact with absent parents. *Journal of Marriage and the Family, 50*(3), 663–677.

Shefts, K. R. (2002). Virtual visitation: the next generation of options for parent-child communication. *Family Law Quarterly, 36*(2), 303–327.

Sillars, A., Canary, D. J., & Tafoya, M. (2004). Communication, conflict, and the quality of family relationships. *Handbook of family communication* (pp. 413–446). Mahwah: Lawrence Erlbaum.

Smyth, B. (2002). Research into parent-child contact after parental separation. *Family Matters, 62*, 33–37.

Stafford, M. (2004). Communication competencies and sociocultural priorities of middle childhood. *Handbook of Family Communication* (pp. 311–332). Mahwah: Lawrence Erlbaum.

U.S. Census of Population and Housing, (2008) *Household relationship and living arrangements of children under 18 years, by age and sex.* Washington: Government Printing Office, 2008. http://www.libraries.iub.edu/index.php?pageId=2558).

Vetere, F., Gibbs, M. R., Kjeldskov, J., et al. (2005). Mediating intimacy: designing technologies to support strong-tie relationships. *Proceedings of the CHI* (pp. 471–480). New York: ACM.

Wallerstein, J. S., & Kelly, J. (1996). *Surviving the breakup: how children and parents cope with divorce.* New York: Basic Books.

Yarosh, S., & Abowd, G. D. (2011). Mediated parent-child contact in work-separated families. *Proceedings of the CHI.* New York: ACM.

Yarosh, S., Chew, Y. C., & Abowd, G. D. (2009a). Supporting parent-child communication in divorced families. *International Journal of Human Computer Studies, 67*(2), 192–203.

Yarosh, S., Cuzzort, S., Müller, H., & Abowd, G. D. (2009b). Developing a media space for remote synchronous parent-child interaction. *Proceedings of the IDC* (pp. 97–105). New York: ACM.

Chapter 6
Kids & Video: Playing with Friends at a Distance

Kori M. Inkpen

Abstract As children's use of technology grows, we see video as an important communication medium for children to connect with their friends and family members. This chapter describes a series of research projects focused on connecting children with their friends using video. The VideoPlaydate project explored children's use of synchronous video conferencing technologies to connect with distant friends and examined several extensions to standard videoconferencing systems to better support children's free play. In a follow-up project called IllumiShare, a novel hardware device was developed to enable any surface to become shared. IllumiShare allows children to easily incorporate any physical object into their remote play with friends, including toys, books, and games. The chapter also describes a project which explored children's use of an asynchronous video messaging tool called VideoPal to help children develop new friendships with Pen Pals from a different country or strengthen existing friendships with children they see on a regular basis. These research projects demonstrate the potential of video to connect children with their peers, and also identifies several important design recommendations that must be considered in systems to support children's remote play with friends.

Introduction

Video is an exciting new medium for children, especially in the ways that video conferencing technology can support children's rich social interactions with friends and family members. Many researchers have explored the potential of video to connect children with distant family members such as grandparents (Follmer et al. 2010; Raffle et al. 2011a, b), and travelling or divorced parents (Yarosh and Abowd 2011); however, video also has huge potential to also support children's interactions

K. M. Inkpen (✉)
Microsoft Research,
1 Microsoft Way, Redmond, WA 98052, USA
e-mail: kori@microsoft.com

C. Neustaedter et al. (eds.), *Connecting Families,*
DOI 10.1007/978-1-4471-4192-1_6,© Springer-Verlag London 2013

with their friends (Yarosh et al. 2010; Yarosh and Kwikkers 2011; Du et al. 2011; Inkpen et al. 2012).

Consumer use of video communication is expected to grow substantially in the coming years, from 600 million video chats in 2008 to just under 30 billion in 2015 (Poor and Wolf 2010). Interestingly, statistics on adults' use of video communication reveals that younger Internet users (ages 18–29) are more likely to use video calls compared to older adults (Rainie and Zickuhr 2011). While there is little data on the growth of video communication for children, children's increasing access to computer technology and their use of rich media could significantly add to the growth of video communication.

Many innovative prototypes have been designed to support children's social play. For example, sharing digital images was explored by Lindley et al. (2010) in a system called Wayve, which enables sharing of handwritten and photo messages to support social interactions within families. Although Wayve was originally designed to help families manage their practical affairs, user studies revealed that it encouraged playful use, particularly for children. Other work by Mäkelä et al. (2000) also showed that leisure sharing of digital images supports playful interactions (joking, expressing emotions, and creating art) to share current activities and feelings.

Connected toys have also been explored to encourage children's free play with remote friends. Bonanni et al. (2006) explored children's play using networked, wireless, robotic figurines called PlayPals. PlayPals consist of two or more dolls that are remotely synchronized such that when one doll is moved the remote doll moves in the same way. There are also tangible tokens that can be placed in the doll's hand to provide additional functionality such as voice and video communication. In a user study the concept of connected toys was very intriguing for the children; it enriched their play and gave them new ways to communicate their thoughts and feelings. However, the dolls alone were not enough—social play only occurred when the children were also provided with a synchronous audio connection. Yarosh and Kwikkers (2011) also recommended the use of remote toy interaction to support children's play. This could involve interaction between remote physical toys, or children's interaction with a virtual representation of a remote physical toy.

For reasons that this chapter will describe, video provides a unique opportunity for children to engage in rich, social play with their friends. In what follows, we explore the potential of synchronous and asynchronous video to support children's communication and play with their friends. These friends could be distant relatives, Pen Pals, or school friends that they see regularly. We first review the potential benefits of video communication for children. We then discuss the use of synchronous video to support children's free play and present results from the Video Playdate (Yarosh et al. 2010) and IllumiShare (Junuzovic et al. 2012) projects. We then present research on children's use of asynchronous video, including results from the VideoPal projects (Du et al. 2011; Inkpen et al. 2012). Overall design recommendations for children's video communication are then presented and finally we close with a discussion of the future potential of connecting children with video.

Video Communication for Kids

One of the key benefits of video is that it supports non-verbal communication such as the use of gestures, body language, facial expressions, and voice expressions (Mehrabian 1972), and can convey emotional signals to eliminate confusion in conversations (Ekman and Friesen 1968). Supporting children's non-verbal communication is important, since children's communication abilities are typically less mature than adults (Piaget 1926). Mediums that leverage actions, body movement or imagery might be easier for children to use than text based communication such as email (Bruner 1975).

Several Computer Mediated Communication (CMC) theories suggest that video could be a desirable medium to facilitate communication among children because of its capabilities in supporting nonverbal communication. According to media richness theory, video allows people to simultaneously observe multiple nonverbal behavioral cues, including body language, facial expression and tone of voice (Daft and Lengel 1984). Social presence theory points to the fact that communicating partners can have more awareness about each other's states using video than other media like email, text messages or over the telephone (Short et al. 1976). Furthermore, common ground theory suggests that enhanced mutual awareness among communicating partners provides grounding necessary for the development of conversations, thereby making communication more effective (Clark and Brennan 1991). The contextual information provided in video therefore suggests that it is a more effective medium for communication than text-based media, like email, IM, or SMS, or voice-based media, like telephone.

There has been a long history of research exploring synchronous Video Mediated Communication (VMC) in the workplace, however, much of the literature has failed to show benefits of video over audio on objective measures such as time to complete a shared task (Kirk et al. 2010; Whittaker 2003). However, studies in the workplace have found that video can enhance verbal descriptions with gestures, convey nonverbal information, express attitudes in posture and facial expression, and manage and interpret pauses, thus making communication more effective (Isaac and Tang 1994). Despite the extensive study of VMC in the workplace, and the plethora of enterprise systems developed over the years, usage continues to be relatively low.

In home settings, the use of video is growing rapidly because of a desire for closeness and has been shown beneficial to support people's desire to stay connected to family members and close friends (Kirk et al. 2010; Romero et al. 2009; Tee et al. 2009). VMC applications like video conferencing and video chat have been used increasingly to connect to extended family members and close friends who are separated by long distances and the potential of this technology has received a great deal of media attention (e.g., Harmon 2008). It has been found that VMC can allow family members and friends to feel more connected, and also enable them to share activities with each other in real time (Kirk et al. 2010; Ames et al. 2010; Judge et al. 2011; Judge et al. 2010). When asked what they meant by feeling "close", participants in the Kirk et al. (2010) study expressed that video helped people know each other better, such as children and their grandparents. It also enables young children to converse more effectively than they can over the telephone. Additionally, people desired video because they wanted to be involved in their family's or friends' ongoing lives, take part in routine activities, and just know that someone is there.

Being able to enhance the feeling of "being there" is one key potential of video communication. Researchers have explored young children's interaction with video communication to see if it could provide similar benefits to having their parent be there physically (Tarasuik et al. 2011). The results of this work demonstrated that young children connecting with their parents over video had similar effects as when the parents were physically present, such as exhibiting a similar level of interactivity in both the video and in-person conditions.

Examining children's use of VMC with adults, several studies have found that synchronous VMC has great potential to help young children and adults feel connected. For example, Ballagas et al. (2009) suggested video-mediated communication may be particularly appropriate for communication with young children because it provides better resources for grounding conversation and supports playfulness in remote communication. Ames et al. (2010) compared young children's use of phones and synchronous video conferencing systems to interact with adults. These children enjoyed video chat more than telephone conversations, and were more engaged with video, which led to longer and richer communication. Also, the visual medium enabled activities that would not have been possible with the phone and the children were able to have different levels of participation in the conversation.

In a study of work-separated families Yarosh and Abowd (2011) also found that in some families video chat was an effective way for children (age 7–13) to stay in touch with a remote parent. Their participants reported that video was more emotionally expressive than phone conversations which led to longer conversations and allowed children to engage in show and tell. Unfortunately, participants also reported barriers that limited their ability to use video: problems with setup overhead, lack of necessary infrastructure such as a computer or reliable connection, and the requirement for dedicated time without being able to multi-task (e.g., washing the dishes while talking on the phone). A few families also used online gaming to maintain contact while apart, but several challenges were encountered including lack of support for multiple players on the same computer (so multiple kids could play with the remote parent), difficulty keeping younger children involved in games, and some children's lack of interest in playing with their parents.

Several researchers have looked at ways of extending video conferencing technology to better support children's play with remote adults. For example, Follmer et al. (2010) explored four design approaches for shared play activities to support family togetherness. These activities involved games and book reading activities in a system called Video Play which augmented traditional videoconferencing. Results from initial trials demonstrated that the activities were engaging to both young children (ages 1–7) and their parents, but that some scaffolding was necessary. One concept from this work, Story Places, was found to be a particularly compelling activity for children to engage in with distant family members. In follow-up work Ballagas et al. (2010) explored a distributed interactive book-reading system to improve the feeling of connectedness for long-distance families. Further studies of this system (renamed StoryVisit) revealed that young children were more engaged in video-chat sessions when an e-book was incorporated (Raffle et al. 2011a, b).

Most video communication technologies have been primarily designed to support conversations, however, families often want to incorporate physical artifacts into their play. Researchers have begun exploring technologies that enable physical objects to be incorporated into play between children and a remote parent. For example, the Virtual Box project (Davis et al. 2007) explored asynchronous remote play by allowing a parent to place a virtual gift box on the floor plan of the child's home that the child could later try to find with the aid of a location sensitive PDA. Yarosh et al. (2009) studied parent-child pairs playing a board game together in a media space that included face-to-face video and a shared tabletop video task space. They found that parents and children were able to socially negotiate rules and access to the physical artifacts in the remote space.

In summary, VMC shows a lot of promise for connecting children with adults since video can support rich cross-generational play. Additionally, children's sense of connection comes more from play than discussion. This suggests that video could be beneficial to support children's remote play with their peers.

Synchronous Video to Support Children's Remote Play

Free play is characterized as an unconstrained activity in which children initiate and direct their own interaction with each other and their environment (Johnson et al. 1987). Time spent in free play is a critical part of a child's cognitive development (Vygotsky 1966) and to developing sociocultural and emotional competencies between infancy and adolescence (Stafford 2004).

Social scientists have been exploring children's play for many decades, from the early investigations of Vygotsky (1966) and Piaget (1926) to the current work of the National Institute for Play (2009). The National Institute for Play identifies seven patterns that constitute the elements of play: (1) attunement play (the interplay of affective feedback such as returning a smile); (2) body play; (3) object play; (4) social play; (5) pretend play; (6) narrative play; and (7) transformative-integrative play. These elements are often combined during free play episodes.

Parten (1932) and Howes (1980) observed that social play between children is characterized by five stages of mutual regard and reciprocity. At the most basic level, children participate in parallel play—activities in proximity to one another, but without engaging in social behavior. At higher stages, children direct social behaviors to one another and respond to the behaviors of their play partners. At the highest level of social play, children engage in a complementary and reciprocal activity that requires both verbal and non-verbal coordination on their parts. During free play children may frequently switch between various types of social play.

There has been research on playing games over synchronous video such as Batcheller et al.'s work (2007) which observed groups of college students playing the social game "Mafia" mediated by a videoconference. They found that playing over videoconferencing was fun for participants, but introduced new challenges in terms of managing attention, signaling to remote partners, and social distance. In other work Mueller et al. (2003) examined a class of prototypes called exertion interfaces which combine projection of full body video and computer vision techniques to allow remote partners to play sport-like games together. They discovered that exertion interfaces have a great potential to create and strengthen social bonds between adult strangers. All of these investigations however asked participants to play games with pre-established rules rather than free play over a videoconference.

The next sections describe two recent projects that used VMC to support children playing with remote friends: Video Playdate and IllumiShare.

Video Playdate: Supporting Children's Free Play with Video

To understand the challenges and opportunities that video can provide for free play Yarosh et al. (2010) first studied children playing together using toys such as action figures and dolls with a standard videoconferencing client (Windows Live Messenger) using two different setups: laptop to laptop; and large screen TV to large screen TV. This preliminary study indicated that free play was possible over videoconferencing, but was limited to short periods of social play interweaved with longer periods of parallel play. Examples of social play included pretending to be TV characters, singing a song together, role playing using dolls, and narrating a scenario using action figures. When using either the laptop or TV, the children struggled to understand several communication asymmetries that videoconferencing presents. For example, children (as well as adults) have a difficult time understanding the field of view of the web camera, and therefore do not always know what is visible to their friend. Additionally, the children did not have a good awareness of appropriate volume levels and had a tendency to talk very loudly. This seemed to be influenced by the fact that their friends looked like they were far away, and therefore they believed that it was necessary to talk loud (or yell) to be heard. The children also had trouble seeing each other's toys clearly.

Comparing the laptop and TV conditions, the researchers observed that the children could understand each other better and paid more attention to their friends in the laptop condition, however, they also had to remain relatively immobile in front of the screen. In the TV setup, the children took the opportunity to move around the space more freely but they were troubled by the amount of pixilation of the video. The TV condition also introduced too much physical distance between the children, causing the children to walk right up to the screen to try and get closer to their friend (see Fig. 6.1).

As a follow-up to this work, Yarosh et al. (2010) investigated four different videoconferencing prototypes, each with different affordances for controlling the

Fig. 6.1 Children playing together via videoconferencing using either laptops or large screen TVs

Free Play using a Laptop

Free Play using a TV

children's view (see Fig. 6.2). The following sections describe each of the prototypes as well as the strengths and weaknesses of each as observed during a user study.

Vanilla Prototype

The Vanilla prototype simulated a high-resolution low-latency videoconference. Figure 6.2a shows the setup including a high resolution webcam (1,280×1,024), microphone and 24″ display of the remote video stream. The smaller screen on the right echoed the image currently being sent to the remote participant. Despite the fact that the basic feature set of this condition was similar to the commercial systems used in the first study, the Vanilla condition was quite effective and the children were engaged while playing in this condition. This prototype was rated

Fig. 6.2 Four video conferencing prototypes tested in the Video Playdate research. **a** *Vanilla prototype*. The small screen shows what the remote participant sees. **b** *Mobile prototype*. Unlocking the small screen activates the camera on the back of the device, allowing the child to control the remote participant's view. **c** *Smart Pan-Tilt-Zoom prototype*. A researcher controls the pan-tilt-zoom camera (*red box*), allowing the child to request different remote views. **d** *Play Rug prototype*. A floor mat is used as the projection surface for a monochrome view of the remote participant's rug

easiest to use, however, visibility was still a problem and the children sometimes had difficulty making sure that their toys were visible to their friends.

Mobile Prototype

The mobile prototype gave the children the ability to control their friend's view with a simple mobile video device (see Fig. 6.2b). The mobile screen consisted of a 7″ monitor with a standard webcam attached to the back, facing away from the viewer.

When the mobile device was picked up, the camera on the back of the device was activated and the child could point it at anything in their environment they wanted to show their friend.

Again, the children were able to easily play with each other using this prototype, however, many of the children considered it to be the most difficult since they had to hold the device while composing their shots. Additionally, when the mobile component was activated, it replaced the face-to-face view which sometimes made it hard for their partner to understand what they were trying to do. The children that used the mobile condition successfully often used a turn-taking strategy to be able to play together (*"first I show my doll, then you show your doll"*). Despite the challenges it presented, several children found the mobile condition to be very compelling and some commented that *"you could literally be where the person was playing!"* Most of the children selected this condition as the most fun and it tied with one of the other conditions for being the most desired condition.

Smart Pan-Tilt-Zoom Prototype

The Smart Pan-Tilt-Zoom (PTZ) prototype used a PTZ camera with a Wizard-of-Oz methodology where the researchers controlled the PTZ camera (see Fig. 6.2c). The children could direct the PTZ camera by giving a verbal command to specify an area of interest, such as (*"zoom in on the toy car"*). If the children did not provide any direction, the researcher manipulated the PTZ camera to keep the children in view as much as possible.

This prototype enabled the children to move freely about the space, have a clear view of their partner, and also be able to focus on the toys when appropriate. Some of the children liked that the camera automatically chose the appropriate view while others enjoyed being able to easily control their view. At times the children had trouble negotiating who should control the view and had to resolve this conflict socially (e.g., *"okay, ask yours to zoom in on the* [toy]*"*) or through planned sequences of views (*"so start out so we can't see them, and then we go here, and then ta-ta-da!"*). They also sometimes wanted to keep an object (or themselves) hidden. For example, some children expressed *"don't look here, I want to do a surprise"*. One negative aspect of this prototype was that the movement of the PTZ camera was sometimes distracting and some children became disengaged from the session and instead played "dodge-the-camera".

Play Rug

The Play Rug prototype used a camera-projector system to provide a shared floor space for the children to play on. A camera suspended above the play rug (see Fig. 6.2d) captured a video stream of the rug surface and transmitted it to the remote projector. The video stream of the remote floor space was projected directly on top of the local floor space and vice versa. Like the PlayTogether system (Wilson and

Robbins 2007), the visual echo problem (i.e., re-projecting artifacts) was resolved by installing IR filters on the overhead cameras. This restricted the video to be only monochrome, but allowed a standard rug to be used rather than a specialized projection surface.

The children saw potential in this technology and often selected it as the one they would most want to have at home. However, there were several challenges with the prototype. First, it was hard for some children to understand the interweaving of the two physical spaces and some were confused when a physical and a virtual object occupied the same space. Additionally, while being able to occupy the same space allowed for some fun physical play (in fact, this condition had the most movement play), this feature also made it difficult for some children to come to an agreement about the interaction between physical toys. For example, two of the children playing with cars could not agree on an interpretation of events (*"It's rolling over you!" "No, it's rolling under me!"*). Finally, the monochrome projection of the remote activity was often too subtle to attract attention and it was hard for the children to see both the screen and the rug at the same time. This led to some missed opportunities for social play.

Overall Feedback Across the Conditions

Overall, although there was a great deal of individual variability, the children were able to successfully play together using all of the prototypes. Though all four prototypes supported social play equally well, different technologies for managing views led to different types of play among the pairs. The shared task space created in the Play Rug setup supported movement and physical activities, such as play fighting and tumbling. The Mobile setup enabled the children to control their partner's view and encouraged turn-taking and narrative play. However, when view control was simplified in the Vanilla and Play Rug setups, the children could devote more cognitive resources to engaging in pretend play. It is also important to examine whether technology should be designed to support natural play, or add to the experience. Aspects of both the Play Rug and the Mobile setups became a part of the children's play instead of just enabling play.

The results from this project demonstrate the potential of supporting children's free play through video, but also highlights challenges that exist for many video-conferencing environments. We briefly present these opportunities and challenges which helped inform the design guidelines presented later in the chapter.

The first challenge deals with managing the visibility (and invisibility) of objects and toys in the space. This includes problems related to resolution and framing play within the camera view. Interestingly, several of the children used the cushions around the play area to establish a stage for their toys that they knew was clearly visible to the other person.

A second challenge stemmed from the lack of peripheral cues, and the fact that children frequently shift attention between individual and mutual activities during free play. For children, a face-to-face view of their partner was key to their social

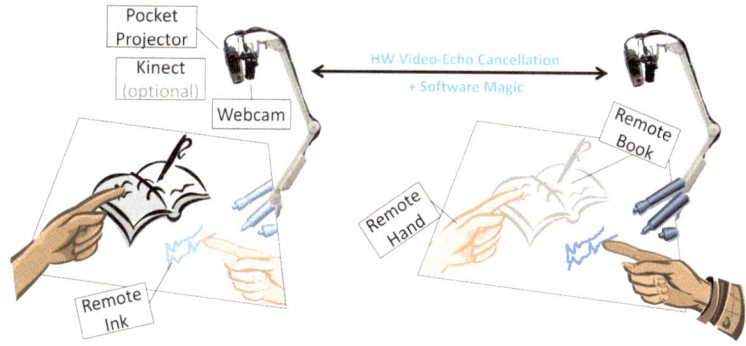

Fig. 6.3 Illustration of table sharing with IllumiShare

play as it was the only reliable clue to the direction of their partner's attention. See-ing their partner attend to their activity led to greater social play, while perception of inattention led the children to play in parallel instead. Managing attention also became more complicated with multiple displays. Elegant view management that both signals the direction of the partner's attention and lets the child appropriately direct their attention is an open challenge for designers.

A third challenge involves helping the children manage intersubjectivity. Inter-subjectivity is defined as the capacity for establishing and maintaining a common ground of engagement among participants involved in an activity together (Winegar and Valsiner 1992). In the context of video-mediated play it involves understanding both what you and your partner see and determining how to act meaningfully to-wards each other. However, play is a cognitively demanding activity that leaves few attention resources available for maintaining a mental model of what the other per-son sees. Children who were most successful at framing their play made frequent use of the feedback screen, but many still seemed to get confused about who sees what.

IllumiShare: Providing a Shared Physical Task Space

Having children be able to easily see and interact with each other's toys is an im-portant part of their play. As shown in the previous section, visibility of toys and children's actions with the toys is often challenging in typical video conferencing environments. Yarosh and Abowd explored this concept for children's interactions with remote adults and developed a system called ShareTable which allows children and their parents to have a shared view of physical artifacts (Yarosh et al. 2009). Junuzovic et al. (2012) designed and built a similar system called IllumiShare which is a cost-effective, light-weight device that enables users to share physical and digital objects on *any* surface while also providing rich referential awareness (see Fig. 6.3). Although IllumiShare is similar to previous devices (e.g., Clearboard,

Ishii and Kobayashi 1992; VideoDraw, Tang and Minneman 1991; PlayTogether, Wilson and Robbins 2007; ShareTable, Yarosh et al. 2009) it enables any surface to be shared, and provides a better quality view of the remote shared space.

IllumiShare enables children to interact with objects in a natural, seamless way, similar to how they would interact in a face-to-face environment, however, their interactions are bounded by the constraints of the system in terms of what can and can't be seen. IllumiShare has a simple affordance—anything in the illuminated area is shared with others. For example, children can draw together on a piece of paper simply by placing the paper underneath IllumiShare. From that point on, they can draw together right on the paper and also see each other's hands as they point at parts of the drawing.

Use of IllumiShare can be combined with a standard videoconferencing session to provide the children with both a face-to-face view of their friend and the shared surface. This is similar to the setup used by Tang et al. (2010) which explored the benefits of providing support for the person-, task- and reference-spaces. Orientation of the shared surface is an issue for all surface sharing systems. Similar to ShareTable, IllumiShare orients the surface in the same direction for both children. This means that the children's hands and arms come out from the same side of the table, as if the children were sitting in the same chair. This also means that the remote-child's hands and arms are disembodied from their front-on view, which is seen across the table. However, consistent with previous research (Tang et al. 2010) the children had no trouble understanding this configuration, and were able to interact naturally.

Junuzovic et al. (2012), studied eight pairs of children (ages 9–11) using IllumiShare during remote play. IllumiShare was combined with a Skype videoconferencing session to support both face-to-face interaction and task-based interaction (see Fig. 6.4). Children played in three different conditions: IllumiShare-only, Video-only; and combined Video+IllumiShare. Audio was provided in all three setups.

The children's play during the IllumiShare sessions was extremely intuitive and the system encouraged natural interaction. They immediately understood the IllumiShare semantics that anything that was lit up by the projector was shared (public) and everything else was private. All of the children understood that if they pointed in the illuminated area, their friend could see their hand, as well as where they were pointing. Interestingly, if a game could not be played remotely with its original rules, the children easily modified the rules.

Overall, the children engaged in 40 different tasks during the play sessions which were clustered into five categories: pen and paper (20); card or dice games (8), showing things (4); gesture games (3); and other games (4). Figure 6.5 shows screenshots from some of the activities. Pen and paper activities consisted of activities such as drawing and writing. Example card or dice games were War or Bowling. Showing things typically involved showing books or magazines. Gesture games were rock/paper/scissors and dancing. The other games included I Spy and Mancala. The pen and paper, as well as dice and card tasks were predominantly performed when IllumiShare was available while gesture games were played when Video was

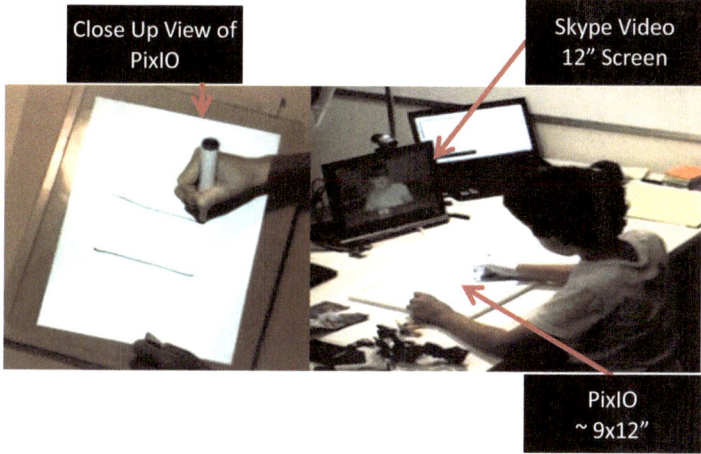

Fig. 6.4 Experimental setup for the IllumiShare user study, which included both Skype video and IllumiShare

Fig. 6.5 Screen shots of children doing various activities with IllumiShare

available. The other games and showing things were mostly performed when both IllumiShare and Video were available.

Video+IllumiShare

The children seemed to thrive in the Video+IllumiShare setup. In all groups, the children were fully engaged as soon as the session started. Often, the first reaction to having both IllumiShare and Video was to write a quick note in the shared area. They also interacted using toys, such as fighting with action figures and arranging toys in playful ways. The children were very animated about what they were doing, even if the task was taking place on the shared surface. For instance, when a pair of boys was playing the card game War, one of them used whole body gestures and as he put cards down. He would say things like "*I summon…an ace!*" in an authoritative wizard like voice as he slammed his card down on the table. Meanwhile, when a group of girls played I Spy, each of them had a copy of the board and when one found an item, she would get extremely excited, put the board into the shared area and point at the item's location. The other would immediately look at the remote board where her friend's hand was pointing in order to find that same location on her own board.

The Video+IllumiShare condition was considered the easiest and most fun. The children explained that it was "*just like being next to them*". When asked which setup they would like to have at home, all but one selected Video+IllumiShare, because "*you can see each other and play on the table*", and "*because you can see the person and see what they are doing*".

Video-Only

When the children had Video but not IllumiShare (i.e., standard video conferencing setup), they seemed to struggle more to play compared to the other conditions. Some were able to adapt quickly, for instance, a pair of girls played I Spy but had to bring the I Spy board up to the camera to point at a location. In other cases the video condition resulted in awkward silence during which the children would glance around the room and look at each other without talking. In one such instance, the silence was broken with "*Oh look, scissors. I can't wait until the table thing works*". Most children ranked the Video condition as being less fun than IllumiShare because "*just video was more of a talk thing. If you wanted to just talk, you would be fine. But if you wanted to play, then video wasn't good*".

IllumiShare-Only

Children performed similar tasks in the IllumiShare and Video+IllumiShare conditions, but they tended to be less visually animated without the video. For instance, the same pair of boys whose game of War was described earlier also played War without the video. In this case, all of the body actions, such as hand motions, were subdued and took place on the shared surface. The absence of video was most noticeable when the children had difficulty interpreting what their friend was doing (for example, if they were not doing anything on the shared task space). In these instances the children would often called out to see if the other person was there and ask what they were doing.

Overall Feedback Across Conditions

IllumiShare had a significant impact on the children's level of engagement during their play. When IllumiShare was removed, engagement decreased while adding IllumiShare back increased engagement. Some children struggled to find something to do without IllumiShare. For example, one girl asked her friend "*What can we do over video chat*" and her friend responded "*I don't know*". The children sometimes reacted negatively to the removal of IllumiShare "*This is bad! This is very, very bad!*" and were excited when it was brought back, "*Oh good*" to "*Yaaaaaay, Table!*" In contrast, the removal or addition of video had little impact on level of engagement.

Overall, combining IllumiShare and Video was extremely compelling in terms of supporting children's remote play. The children's interactions were seamless and natural and the children enjoyed playing together using these technologies.

Playing Together with Asynchronous Video

Although synchronous video is an effective way to connect children with their peers, there are several challenges as well. One of the biggest obstacles is the fact that synchronous video requires both children to be available at the same time. This is problematic for two reasons. First, families are busy and schedules can make it hard to coordinate times for children to connect. This was observed by Modlitba and Schmandt (2008) who studied children's interactions with travelling parents and found that although children prefer using video chat, their parents' busy schedules made it hard to coordinate synchronous video chats. Second, children often do not have any awareness of when their friends are available to connect over video. Unlike the workplace where people spend many hours sitting in front of their computers, children's use of computers in the home tends to be for short periods of time, and can be sporadic. Without having some sort of explicit coordination, it is easy to imagine children missing out on opportunities to connect with their friends.

Using asynchronous video as a more flexible means of connecting families was proposed in work by Cao et al. (2010). In other work, Zuckerman and Maes (2005) proposed the Contextual Asynchronous System (CASY), which enabled family members to send 'good morning' and 'good night' asynchronous video snippets into a shared family database. The recipient could then view the snippet in the context of going to sleep or waking up. An initial prototype of this system found that the asynchronous video snippets increased participants' feeling of connectedness.

Raffle et al. (2011a, b) explored the viability of asynchronous photographic and video messaging for pre-school aged children to communicate with distant relatives. They developed three innovative prototypes that explored a jack-in-the box toy with an embedded mobile phone to enable children to compose and share electronic media. The prototypes work by placing a mobile phone into the Toaster prototype and pressing down which causes the phone to start playing the Pop Goes the Weasel song. While depressed, the phone can take a photo, cue up a video, or display an image on the screen. When the song is done, the phone pops up and displays the media to child. The children's images or performances with the device are automatically captured by the front-facing camera on the phone, and are then shared with remote family members. The *Orange Toaster* took photos of the children; the *Family Toast* device enabled children to use tangible objects to select and browse family photos; and the *Play with Elmo* prototype played videos created by a remote

Fig. 6.6 VideoPal screen shot. The *bottom* half of the screen displays a list of active video conversations and meta-data about those conversations. The *top* part of the screen shows a visualization for one of the conversations

family member. Although the communication aspects of these prototypes have not been extensively studied, this work shows potential for asynchronous messaging to support young children's interactions.

The next section describes recent work exploring children's use of VideoPal, an asynchronous video messaging system to support children's communication with their friends.

VideoPal

VideoPal is an asynchronous video mediated communication tool designed to enable children to easily exchange video messages with their friends to engage in a rich conversation. VideoPal captures video using either a webcam, recording the screen (with or without a voice overlay), or uploading an existing video. Video messages can be sent to one or more friends and are organized by conversation topic to show the flow of a conversation, indicating who responded to whom and when (see Figs. 6.6 and 6.7).

VideoPal was initially used as an educational Pen Pal tool to support the development of cross-cultural friendships (Du et al. 2011). Thirty, 9–12 year old children (15 girls, 15 boys) from the United States and Greece corresponded with each

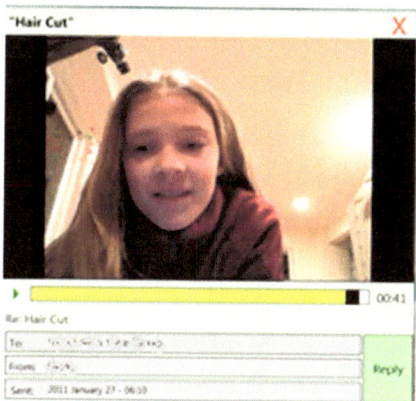 

Fig. 6.7 User Interface to enable children to play and reply to video messages

other using both Email and VideoPal. Results from this work demonstrated that the children preferred VideoPal over Email because it was more fun, it enabled them to get to know each other better, and made them feel closer to their new friends. Furthermore, the children liked VideoPal because it enabled natural communication including speech, body language and facial expressions. These results are consistent with media richness (Daft and Lengel 1984) and social presence (Short et al. 1976) theories and demonstrate that the benefits of synchronous video communication can also be realized with asynchronous video.

VideoPal was also used to examine how asynchronous video could augment children's existing friendships (Inkpen et al. 2012). Just as text messaging has become an important part of youth's social communication (Rideout et al. 2010) video can provide even more richness and enable children to interact with each other in new ways. A 9-week field study was conducted with a group of six girls who used VideoPal in their own homes. The girls, age 9–11, were very close friends and saw each other almost daily.

The girls' usage of VideoPal was overwhelming. Within the first 24 h (which occurred during the girls' school holiday) the girls sent each other 197 video messages. Within the first 2 weeks of the study, 585 messages were exchanged in 93 different conversations. Most of the messages were webcam messages (90 %), and most were sent to all of the girls in the group (60 %). The length of the conversations varied widely, with some conversations only having one message, and others having upwards of 140 messages. Most of the messages were relatively short, with 75 % of them being less than 30 sec. long. Besides just creating messages, the girls received a lot of enjoyment from watching their friends' video messages (as well as their own). During the first 2 weeks of the study, there were 2,670 message views and some messages were viewed upwards of 36 times. When asked what they liked best about VideoPal their responses included because you can *"see your friends"*, *"being able to chat with your friends when they are not with you"*, *"see people's videos even if they're not online"*, and *"send videos when other people aren't on the computer"*.

Although VideoPal was designed as a conversation tool, it was used for much more than just talking. The breadth of use was fascinating and included many types of sharing and play. The videos were coded and clustered into six different genres: conversations; show and tell; sharing activities; screen recording; play acting/performing; and just for fun. The next sections describe each of these genres to show the power video has to connect close friends. Figure 6.8 gives an illustrative example for each genre.

Conversations

Despite the fact that all conversations were asynchronous, there were many videos where the girls would just turn the webcam on and talk to their friends, even though their friends were not actually there. The girls were very comfortable talking over video, and the videos seemed fairly spontaneous, and not rehearsed or planned. The dialog was very conversational as the girls addressed each other, and responded to each other's comments. Many of the conversation videos were normal, everyday exchanges about the things going on in their lives, like homework and what they were doing. Often, the girls' behaviour in the videos seemed as if they were actually talking to their friends face-to-face. They also took advantage of the visual nature of the video medium to aid the conversation when needed.

Show and Tell

The girls liked to create videos to show each other things such as their favourite Christmas presents, their pets, their rock collections, and tours of their rooms. The girls used the mobility of the laptop to walk around their homes and share many different things and they would often show themselves along with the artifacts they were sharing. These show and tell activities were sometimes challenging however, because of problems capturing the artifacts. For example, walking with the camera resulted in too much movement, causing the video to be very jumpy and difficult to watch. It was also awkward to walk around carrying the laptop in one hand, and using the other hand to point the webcam at the items of interest. And even if the girls were more stationary, it was sometimes difficult to position the web camera appropriately to capture the desired scene.

Sharing Activities

Often the girls wanted to be able to share the activities they were currently engaged in, even if their friends were not available. This is consistent with Judge and Neustaedter's (2010) work on video conferencing in the home which demonstrated that families with children primarily used video conferencing to share activities instead of just conversations. For example, the girls created videos of themselves playing

A.   Conversation

"Hey Miki, guess what I had for dinner today, CEREAL! ... I had Lucky Charms ..."

"Um, you always have cereal Hannah, I am like so not surprised ..."

"Miki, you are right, I ALWAYS have cereal ..."

B.   Show and Tell

(singing) "I got something awesome ... A PHONE! .... It's my own phone! Do you want to see some pictures on it?"

C.   Shared Experiences

"Ok guys. I am going to show you my made up beam routine."

D.   Screen Recording

"Hi guys, this is my slide show of Funny Bunnies. So right here is a picture of a bunny popping out of an Easter egg ... "

E.   Play Acting / Performing

Lady Gaga & Beyonce. Telephone Music Video.

F.   Just for Fun

"Watch us roll in our money. WHOOO! <lots of screaming and laughing>"

Fig. 6.8 Example video messages for each of the conversation genres

things like Xbox Kinect, doing gymnastics, and building a playhouse. Sharing activities was quite different than conversations, because they tended to capture larger spaces, such as a whole room, or a full-body view. This was somewhat problematic because today's web cameras are optimized for up-close interaction and typically do not have appropriate zoom levels. Additionally, being able to see the feedback window from a distance was hard, so it was difficult to know what was in view of the camera. Finally, when sharing activities, the girls often moved around a lot, again, making it difficult for the camera to capture.

Screen Recording Videos

Although the screen recording feature was only used 10 % of the time, all of the girls commented that they enjoyed making screen recording videos and liked having this feature. Common uses of the screen recording feature involved narrating slideshows and poems, showing excerpts from online games, and showing YouTube videos. Overall, the girls expressed that this was an important feature in the system and that they liked to be able to share things happening on their screen. However, the user interface for this feature was a little awkward to use, which may have impacted the overall use. The voice overlay feature was also important and was used extensively as almost every screen recording had an associated voice overlay. One of the girls was able to carefully arrange her windows to provide a picture-in-picture experience, showing her face, actions, and gestures along with the screen recording. Several of the girls wanted to be able to create these picture-in-picture style videos.

Play Acting/Performing

There were many videos where the sole purpose was to perform instead of converse. The girls acted out things like scenes from Harry Potter or created lip-synced music videos. To add theatrical effects the girls often used props and sometimes moved in and out of the view of the camera. Some of these videos are similar to the types of things children like to share on YouTube; however, VideoPal enabled them to share their videos securely, with just their close friends. Additionally, instead of being a stand-alone YouTube video, they were often part of a conversation thread, where their friends could provide video replies.

In some of the play acting conversations, the girls' play would follow on from one another, which was refered to as asynchronous play. Similar to how children build off of each other's play activities when face-to-face, there were several conversations where one girl would do something, and others would follow along without any explicit coordination. For example, several girls added videos to a Harry Potter conversation where they each acted out different scenes. This created a storytelling style of play, similar to the types of interactions reported for StoryMat (Cassell and Ryokai 2001).

Just for Fun Videos

Often when children get together face-to-face, they like to do crazy things, just for fun. Many of the girls' conversations fit this characterization. Ludic actions that had no specific purpose, other than to share something fun with their friends such as two girls rolling in play money, a girl throwing candies up in the air and catching them in her mouth, and girls making faces in the camera. VideoPal enabled the girls to do silly things to make their friends laugh, even though their friends wouldn't see the video until later. The girls commented that these types of activities were fun when they were at home alone, and were bored. The girls were also observed creating these types of videos when they were physically together with their friends to support their copresent play, although they still enjoyed posting them on VideoPal to share with the rest of the group.

Summary

This research clearly demonstrates that asynchronous videos can support rich conversations, and that it is an effective way for children to connect with their friends, even when their friends are not available. Video adds richness to the communication not possible in current text media. The standard use of smiley faces and emoticons in text-based communication pales in comparison to the expressiveness in the girls' facial expressions, actions, gestures, and voices. Children have no trouble conversing over asynchronous video and these exchanges can be as natural as face-to-face interactions. Additionally, asynchronous video is beneficial for more than just conversations and can enable children to share many different types of experiences with their friends.

Both boys and girls were equally enthusiastic about VideoPal in the school study and both enjoyed sending and receiving videos. However, it is important to note that the more in-depth, 9-week study only involved girls' use of VideoPal. Although the school data indicates that boys are interested in asynchronous video messaging, it is possible that their use of the system and the content they would share could be quite different than what was observed with the girls. More research is needed to better understand how other factors such as gender and age impact children's use of video when communicating with their friends.

Design Recommendations

Examining results from this body of work provides several guidelines for video-based systems to connect children and support their rich social play.

Camera Control and Framing

For both synchronous and asynchronous video communication, one of the biggest challenges is capturing an appropriate view from the camera. This especially problematic for children's play given how much they move around while playing and the fact that they often want to share a large play area. One possible solution is to provide some automatic, or user-guided camera control, where the children can specify what should be in view of the camera, and then have the camera automatically capture the scene by tracking objects or people and panning, zooming, and cropping accordingly. The objects being tracked could be the children themselves, the toys they are playing with, or other markers in the scene that children use to delineate an area. Although there were some concerns with the automatic camera approach in the Video Playdate study, this method would give the children more control over what is captured, and what the camera follows.

A second tension surrounding camera control and framing is ensuring that the children maintain an awareness of what is being captured and shared with their friends. If markers are being placed in the scene as suggested above, this could provide cues to the children about what is visible (as well as what is not visible). However, if the camera is performing more complex pan and zoom operations, some sort of feedback window will be necessary to show what the camera is capturing. Ideally, this feedback window should be positioned in such a way that it is easy for the children to see and does not distract from their activities. For example, the current design of IllumiShare provides a natural affordance of what is being shared given that the projector illuminates the shared space, making it easy for the children to understand what their friends can see (anything placed in the illuminated area), and cannot see (anything outside the illuminated area).

Multiple Camera Streams

Many synchronous video conferencing applications are moving towards transmission of multiple video streams to support group-based videoconferencing. Support for children's play will also benefit greatly from capture and transmission of multiple camera streams. Depending on the type of play, children often want to show their own image, as well as toys or artifacts in their environment, or a screen recording of a game or virtual world they are playing in. Providing multiple video streams enables children to share richer context. Previous work by Gaver et al. (1993) also suggested that three types of views (face-to-face video, task space video, and room context video) were useful, but that switching between video views was challenging because it undermined a person's ability to know what their remote partner was looking at or could see at any one time. Multiple video streams could also be used to better facilitate group play, but the design of the system would need to attend to the issue of intersubjectivity and the ability for each person in the group to know what the others are seeing.

Mobility

As evidenced in much of the previous work, mobility is important for children's play. This is consistent with the results from Judge and Neustaedter's research on video conferencing in the home (2010) which showed that families with laptops would move them around to share activities from different locations in the home. Children's play is rarely restricted to one specific location, and even during play, children may move from place to place. As such, technology to facilitate children's play should be flexible enough to support mobility. Laptop computers and tablets provide some mobility for video conferencing by enabling children to take the device into any room in their home, however, the form factor still makes it awkward to carry from place to place, and movement during play is problematic often resulting in jumpy video that is difficult to follow.

The form factor of a mobile phone may be better for scenarios where mobility is important, particularly when capturing video outside of the home. However, while mobile phones are more conducive to moving around, the small screen may be restricting to the children's experience. First, it would be hard to see the friends they are playing with, as well as get feedback on the video they are sharing. Second, the mobile phone often needs to be held, making hands-free use difficult. Future work exploring different form factors for video capture and playback de-vices is needed to better understand ways to enable mobility while still providing a rich, engaging experience for the children.

Blur Temporal Boundaries

Synchronous video enables children to connect in a rich, face-to-face-like manner, however, scheduling and coordination can be a problem. Asynchronous video helps overcome these issues and enables children to connect with their friends at any time. In the VideoPal studies, some of the messages were exchanged when the children were online at the same time, and as such, these messages were more analogous to rapid-asynchronous exchanges (i.e., when all parties are online at the same time and messages are exchanges in a more synchronous manner). In these situations, the children would often prefer to connect using synchronous video. We see potential for both synchronous and asynchronous video to support children's play, and ultimately, a system that enables seamless shifting between synchronous and asynchronous modes of video communication could provide the best of both worlds.

Ease of Use Critical

A critical issue for video communication systems is ease of use. Too many existing video communication systems require substantial overhead to setup a video call or require that all users have the same software system. This limits how frequently people will choose to utilize the system, as well as whom they are able to talk to. One of the successes of the VideoPal system was how easy it was for the children to use. Although, a great deal of the functionality in VideoPal exists in other software (e.g., video messages can be recorded using webcam software and attached to an email message), VideoPal streamlined the process and made it very easy for children to use. This helped encourage extensive use of the system.

Privacy & Security

Sharing videos publically or with a group of friends has become commonplace with systems such as YouTube, Vimeo and Facebook (Moore 2011). However, many of these videos are broadcast in nature and don't reflect a back-and-forth conversation. Using video for a conversation, or to play with friends is a more personal exchange, and as such, privacy and security issues are important. If the goal is to support children's play, privacy and security becomes extremely important to ensure that the videos are only available for the intended audience, and that the children's safety is ensured. Appropriate parental controls and monitoring must be provided.

Another challenge for video communication in the home is the fact that several different family members may be using the same system to communicate with different people. However, unlike office scenarios, family members are often comfortable with a higher level of sharing, and prefer fast, easy access instead of cumbersome log-off/long-on procedures. A more nuanced approach to family accounts is likely needed to support individual and family video communication (Egelman et al. 2008).

Video Search

Although video is an extremely rich communication medium, it can be difficult to index and search. For example, the VideoPal system uses one frame of the video as a thumbnail for the message; however, because many of the videos start out as a "talking-head" video, most of the thumbnails look alike. This makes it very difficult to find a particular video. One possible alternative is to use speech-to-text systems to automatically record the words spoken, and enable users to search the transcripts. Although this is feasible in theory, it is extremely difficult to do in the context of children's play since children's voices are challenging for automatic transcription (Potamianos 2003). Additionally, the expressiveness of the children's voices (e.g., excitement, enthusiasm) makes this problem even more complex. More work is needed to provide better ways to index and search video content.

Social Networking

The asynchronous nature of VideoPal meant that the content was archived and could be shared with a group (if desired). Sixty percent of the VideoPal videos were shared with the entire group of six girls and 73 % were shared with more than one person. This type of sharing is missing from synchronous video exchanges. Providing group, social networking experiences, even within a closed group is beneficial to help foster common ground between the members and help build a stronger sense of community. Providing these types of benefits for synchronous video communication would also be beneficial and should be explored in future work.

Offline Awareness

The extended VideoPal study was successful in part because the girls were given their own laptop computers and they spent a great deal of time using the laptops. A more common scenario would be a family having a shared, family computer that the children use from time-to-time, resulting in sporadic (and potentially infrequent) use of the computer. In this scenario, awareness of when the children have new video messages, or when their friends are available for synchronous play, would be extremely beneficial. Offline awareness could be provided through objects such as a mobile phone or a toy.

Conclusion

In summary, previous research had clearly demonstrated that video is a rich medium for children which can be used to support children's play. As the presence of consumer videoconferencing in the home grows, video becomes a viable medium to connect children who are both near and far. Whether it is a quick 5 min conversation, or a 2 h playdate, children enjoy engaging with their friends over video. However, as shown through the work presented here, there is no one perfect system. There are many different types of activities that children want to engage in, within many different contexts. Additionally, children's capabilities and desires can differ greatly with age which will impact which systems are most appropriate. For example, 5-year old boys that want to play together with action figures will have different needs than 13-year old girls playing a board game. Better understanding of the types of activities children want to engage in, as well as the advantages and disadvantages of different technological approaches will help inform the design of distributed-play devices.

One of the most striking observations from the research presented in this chapter was the children's level of comfort with video, and their strong desire to engage with their friends using rich media. We see children as potential media trendsetters when it comes to video communication. Previous generations of youth heavily utilized text messaging as their key communication medium. The next generation of kids will likely leverage the richness of video to communicate and play with their friends. Although more research is needed to better understand the best ways to support children's play over video, we strongly believe that this is the way children (and adults) will regularly communicate in the future.

Enabling children to engage in remote play with their friends represents a new usage model of video. Video is traditionally used to connect people who live far away and don't have an opportunity to interact face-to-face. Much of the research in this chapter is concerned with connecting close friends in a way that augments their existing face-to-face relationship. Just as text-messaging has become a dominant way to interact with close friends, video could also enhance existing relationships. Finally, the children also demonstrated a strong desire to share more than just a "talking head". This suggests the need for video communication to move beyond just conversations, to the sharing of rich experiences.

Acknowledgements I would like to acknowledge all of the colleagues who helped envision this space and who worked on the Video Playdate, IllumiShare, and VideoPal studies, as well as others who provide important assistance along the way. A.J. Brush, Svetlana Yarosh, Honglu Du, Sasa Junuzovic, Aaron Hoff, Paul Johns, John Tang, Asta Roseway, Konstantinos Chorianopoulos, Mary Czerwinski, Tom Gross, Gina Venolia, Danyel Fisher, Michail Giannakos, Anoop Gupta, Tom Blank, Zhengyou Zhang, Rajesh Hegde, Brian Meyers, Mike Sinclair, Hrvoje Benko, Andy Wilson, Sarah Morlidge, Holly Senaga, Christi Taylor, Rosa Chang, Olga Lymberis, Scott Saponas, Cati Boulanger, and Peter Ljungstrand. And a special thanks to all of the children who participated in our studies, especially Gabi, Katie, Hayley, Miyeko, Anna, Mia, and Declan.

References

Ames, M. G., Go, J., Kaye, J. J., & Spasojevic, M. (2010). Making love in the network closet: the benefits and work of family video chat. *Proceedings of the CSCW 2010* (pp. 145–154). New York: ACM

Ballagas, R., Kaye, J., Ames, M., Go, J., & Raffle, H. (2009). Family communication: phone conversations with children. *Proceedings of the IDC 2009* (pp. 321–324). New York: ACM

Ballagas, R., Raffle H., Go, J., Revelle, G., Kaye, J., Ames, M., Horii, H., Mori, K., & Spasojevic, M. (2010). Story time for the twenty-first century. *IEEE Pervasive Computing, 9*(3), 28–36.

Batcheller, A. L., Hilligoss, B., Nam, K., Rader, E., Rey-Babarro, M., & Zhou, X. (2007). Testing the technology: playing games with video conferencing. *Proceedings of the CHI 2007* (pp. 849–852). New York: ACM.

Bonanni, L., Vaucelle, C., Lieberman, J., & Zuckerman, O. (2006). PlayPals: tangible interfaces for remote communication and play. *Extended Abstracts of CHI 2006* (pp. 574–579). New York: ACM.

Bruner, J. S. (1975). The ontogenesis of speech acts. *Journal of Children Language, 2*, 1–40.

Cao, X., Sellen, A., Brush, A. J., Kirk, D., Edge, D., & Ding, X. (2010). Understanding family communication across time zones. *Proceedings of the CSCW 2010* (pp. 155–158). New York: ACM

Cassell, J., & Ryokai, K. (2001). Making space for voice: technologies to support children's fantasy and storytelling. *Personal and Ubiquitous Computing, 5*(3), 169–190.

Clark, H. H., & Brennan, S. E. (1991). *Grounding in communication. Perspectives on social shared cognition.* Washington, DC: American Psychological Association.

Daft, R., & Lengel, R. (1984). Information richness: a new approach to managerial behavior and organization design. In B. Staw & L. L. Cummings (Eds.), *Research in organizational behavior* (pp. 191–233). Greenwich: JAI Press.

Davis, H., Skov, M. B., Stougaard, M., & Vetere, F. (2007). Virtual box: supporting mediated family intimacy through virtual and physical play. *Proceedings of the OzCHI 2007* (pp. 151–159). Adelaide: ACM

Du, H., Inkpen, K., Chorianopoulos, K., Czerwinski, M., Johns, P., Hoff, A., Roseway, A., Morlidge, S., Tang, J., & Gross, T. (2011). VideoPal: exploring asynchronous video-messaging to enable cross-cultural friendships. *Proceedings of the ECSCW 2011* (pp. 273–292). Heidelberg: Springer

Egelman, S., Brush, A. J., & Inkpen, K. (2008). Family accounts: a new paradigm for user accounts within the home environment. *Proceedings of the CSCW 2008* (pp. 669–678). New York: ACM

Ekman, P., & Friesen, W. (1968). Nonverbal behavior in psychotherapy research. *Research in Psychotherapy: Proceeding of the Third Conference, 3,* 179–183.

Follmer, S., Raffle, H., Go, J., Ballagas, R., & Ishii, H. (2010). Video play: playful interactions in video conferencing for long-distance families with young children. *Proceedings of the IDC 2010* (pp. 49–57). New York: ACM

Gaver, W. W., Sellen, A., Heath, C., & Luff, P. (1993). One is not enough: multiple views in a media space. *Proceedings of the INTERACT 1993 and CHI 1993* (pp. 335–341). New York: ACM

Harmon, A. (2008). Grandma's on the computer screen. *The New York Times.* November 28, 2007, p. A1.

Howes, C. (1980). Peer play scale as an index of complexity of peer interaction. *Developmental Psychology, 16,* 371–372.

Inkpen, K., Du, H., Hoff, A., Johns, P., & Roseway, A. (2012). Video kids: Augmenting close friendships with asynchronous video conversations in VideoPal. *Proceedings of the CHI* (pp. 2387–2396). New York: ACM

Isaac, E. A., & Tang, J. (1994). What video can and cannot do for collaboration: a case study. *Multimedia Systems, 2*(2), 63–73.

Ishii, H., & Kobayashi, M. (1992). Clearboard: a seamless medium for shared drawing and conversation with eye contact. *Proceedings of the CHI 1992* (pp. 525–532). Monterey: ACM

Johnson, J. E., Christie, J. F., & Yawkey, T. D. (1987). *Play and early childhood development.* Glenview: Scott, Foresman and Company.

Judge, T. K., & Neustaedter, C. (2010). Sharing conversation and sharing life: video conferencing in the home. *Proceedings of the CHI 2010* (pp. 655–658). Montreal: ACM

Judge, T. K., Neustaedter, C., & Kurtz, A. F. (2010). The family window: the design and evaluation of a domestic media space. *Proceedings of the CHI 2010* (pp. 2361–2370). New York: ACM

Judge, T. K., Neustaedter, C., Harrison, S., & Blose, A. (2011). Family portals: connecting families through a multifamily media space. *Proceedings of the CHI 2011* (pp. 1205–1214). Vancouver: ACM

Junuzovic, S., Inkpen, K., Blank, T., & Gupta, A. (2012). IllumiShare: sharing any surface. *Proceedings of the CHI 2012* (pp. 1919–1928). New York: ACM.

Kirk, D. S., Sellen, A., & Cao, X. (2010). Home video communication: mediating 'Closeness'. *Proceedings of the CSCW 2010* (pp. 135–144). New York: ACM

Lindley, S. E., Harper, R., & Sellen, A. (2010). Designing a technological playground: a field study of the emergence of play in household messaging. *Proceedings of the CHI 2010* (pp. 2351–2360). New York: ACM

Mäkelä, A., Giller, V., Tscheligi M., & Sefelin, R. (2000). Joking, storytelling, artsharing, expressing affection: A field trial of how children and their social network communicate with digital images in leisure time. *Proceedings of the CHI 2000* (pp. 548–555). New York: ACM

Mehrabian, A. (1972). *Nonverbal communication.* Chicago: Aldine-Atherton.

Modlitba, P., & Schmandt, C. (2008). Globetoddler: designing for remote interaction between preschoolers and their traveling parents. *Extended Abstracts of CHI 2008* (pp. 3057–3062). New York: ACM.

Moore, K. (2011). 71 % of online adults now use video-sharing sites. Pew Internet & American Life Project, 7/25/2011. http://pewinternet.org/Reports/2011/Video-sharing-sites.aspx. Accessed 21 Sept 2011.

Mueller, F., Agamanolis, S., & Picard, R. (2003). Exertion interfaces: sports over a distance for social bonding and fun. *Proceedings of the CHI 2003* (pp. 561–568). New York: ACM

National Institute for Play. (2009). *Play science: the patterns of play.* Carmel Valley.

Parten, M. B. (1932). Social participation among preschool children. *Journal of Abnormal and Social Psychology, 27,* 243–269.

Piaget, J. (1926). *The language and thought of the child.* New York: Harcourt, Brace & Company.

Poor, A., & Wolf, M. (2010). Report: the consumer video chat market, 2010–2015. GigaOM Pro, June 7, 2010.

Potamianos, A. (2003). Robust recognition of children's speech. *IEEE Transactions on Speech and Audio Processing, 11*(6), 603–616.

Raffle, H., Ballagas, R., Revelle, G., Mori, K., Horii, H., Paretti, C., & Spasojevic, M. (2011a). Pop goes the cell phone: asynchronous messaging for preschoolers. *Proceedings of the Interaction Design and Children* (IDC 2011) (pp. 99–108). New York: ACM.

Raffle, H., Revelle, G., Mori, K., Ballagas, R., Buza, K., Horii, H., Kaye, J. Cook, K., Freed, N., Go, J., & Spasojevic, M. (2011b). Hello, is grandma there? Let's read! StoryVisit: family video chat and connected E-Books. *Proceedings of the CHI 2011* (pp. 1195–1204). New York: ACM.

Rainie, L., & Zickuhr, K. Video calling and video chat. Pew Internet & American Life Project. http://pewinternet.org/Reports/2010/Video-chat.aspx. Accessed 26 Nov 2011.

Rideout, V. J., Foehr, U. G., & Roberts, D. F. (2010). Generation M2: Media in the lives of 8- to 18-Year-Olds. A Kaiser Family Foundation Study, January 2010. http://www.kff.org/entmedia/upload/8010.pdf. Accessed 21 Sept 2011.

Romero, N., Markopoulos, P., Baren, J., van, Ruyter, B., de, Jsselsteijn, W., & Frashchian, B. (2009). Connecting the family with awareness systems. *Personal and Ubiquitous Computing, 11,* 303–329.

Short, J., Williams, E., & Christie, B. (1976). The social psychology of telecommunication. London: Wiley.

Stafford, M. (2004). Communication competencies and sociocultural priorities of middle childhood. *Handbook of family communication* (pp. 311–332). Mahwah: Lawrence Erlbaum.

Tang, J. C., & Minneman, S. L. (1991). Videodraw: a video interface for collaborative drawing. *Transactions on Information Systems, 9*(2), 170–184.

Tang, A., Pahud, M., Inkpen, K., Benko, H., Tang, J. C., & Buxton, W. (2010). Three's company: understanding communication channels in three-way distributed collaboration. *Proceedings of the CSCW 2010* (pp. 271–280). New York: ACM.

Tarasuik, J. C., Galligan, R., & Kaufman, J. (2011). Almost being there: video communication with young children. *PLoS One, 6*(2), e17129. doi:10.1371/journal.pone.0017129.

Tee, K., Brush, A. J., & Inkpen, K. M. (2009). Exploring communication and sharing between extended families. *International Journal of Human-Computer Studies, 67*(2), 128–138.

Vygotsky, L. (1966). Play and its role in the mental development of the child. *Voprosy Psikhologii, 6.*

Whittaker, S. (2003). Things to talk about when talking about things. *Human-Computer Interaction, 18,* 149–170.

Wilson, A. D., & Robbins, D. C. (2007). PlayTogether: playing games across multiple Interactive tabletops. *IUI Workshop on Tangible Play: Research and Design for Tangible and Tabletop Games, IUI 2007,* pp. 53–56.

Winegar, L. T., & Valsiner, J. (1992). *Children's development within social context.* Mahwah: Lawrence Erlbaum.

Yarosh, S., & Abowd, G. D. (2011). Mediated parent-child contact in work-separated families. *Proceedings of the CHI 2011* (pp. 1185–1194). New York: ACM.

Yarosh, S., & Kwikkers, M. R. (2011). Supporting pretend and narrative play over videochat. *Proceedings of the IDC 2011* (pp. 217–220). New York: ACM.

Yarosh, S., Cuzzort, S., Müller, H., & Abowd, G. D. (2009). Developing a media space for remote synchronous parent-child interaction. *Proceedings of the IDC 2009* (pp. 97–105). New York: ACM.

Yarosh, S., Inkpen, K. M., & Brush A. J. (2010). Video Playdate: toward free play across distance. *Proceedings of the CHI 2010* (pp. 1251–1260). New York: ACM.

Zuckerman, O., & Maes, P. (2005). Awareness system for children in families. *Proceedings of the IDC 2005*. Poster. New York: ACM.

Part III
The Extended, Distributed Family

Chapter 7
Connecting Families across Time Zones

Xiang Cao

Abstract Nowadays it has become increasingly common for family members to be distributed in different time zones. These time differences pose specific challenges for communication within the family and result in different communication practices to cope with them. This chapter discusses these challenges and practices based on a series of interviews with people who regularly communicate with immediate family members living in other time zones. We found that families rely on synchronous communication despite the time difference, implicitly coordinate their communication through soft routines, and show their sensitivity to time in various forms. These findings allow us to reflect on the meanings of time difference in connecting families, and design opportunities for improving the experience of such cross time zone family communication.

Introduction

The last century has seen vast advances in both transportation and communication technologies, shrinking the world into a "global village". As a result, not only do people more frequently travel and communicate internationally in work settings, but it is also increasingly common for members of the same family to be living in different regions, countries, or even continents. For example, grown-up children leave home to study abroad; spouses work for companies in distant locations; siblings pursue different life paths around the world and so on. Communication tools such as telephone and email can in some sense render the distance irrelevant—reaching your family halfway around the world can be just as immediate as if they were living in the same city. The recent prevalence of internet technology has also made such long-distance communication accessible and affordable on a daily basis.

X. Cao (✉)
Microsoft Research Asia,
Beijing, China
e-mail: xiangc@microsoft.com

C. Neustaedter et al. (eds.), *Connecting Families,*
DOI 10.1007/978-1-4471-4192-1_7, © Springer-Verlag London 2013

Remote family members had never had as many ways to communicate as they do today.

However, modern communication technologies also highlight one specific factor in long-distance family communication that was once negligible—the time difference. Being geographically far away from each other often also means the family members are living in different time zones. In the previous era when such long-distance communication was dominated by asynchronous channels such as letter, time difference was essentially "transparent" since the time taken to deliver the message itself was often much greater than the time difference. Yet as people become accustomed to contemporary telecommunication technologies and start to expect immediacy and synchronicity for all communication with family, time difference "suddenly" comes into play. Calling your parents becomes tricky when their day is your night; text messages to loved ones might be read half a day later; and when you have something exciting to share with your family, there is simply nobody awake to hear about it. This "time distance" seems to pose more challenges than geographical distance for communication between today's remote family members. Understanding these challenges, as well as how people currently deal with them, not only provides a special lens into the broader scene of family connection, but can also guide us to design better communication tools to suit the special needs of families living across time zones.

Indeed, recent investigations on family communication have already identified time difference as an important factor. For example, Modlitba and Schmandt (2008) found that parents travelling to other time zones adjust their schedule to suit the bedtime of children at home. In the study of BuddyClock (Kim et al. 2008), a device that shares sleeping status between family members, time difference was reported as a potential reason why such information would be needed. Lottridge et al. (2009) reported remote couples taking time differences into account to predict the partner's availability and whereabouts. Time differences can also cause behavior changes. Lindley et al. (2009) reported that time difference was one of the challenges that contributed to older adults' adoption of asynchronous communication methods such as email.

Obviously, the influences of time difference spread much beyond the realm of family communication. Zerubavel (1985) discussed the social impact of time and time zones in general, and more specifically, Tang et al. (2011) investigated how globally distributed work teams collaborate across time zones. Indeed, practice around time differences in work settings constitutes an interesting counterpart to that in the family setting, revealing different values in the two environments.

The rest of the chapter will discuss findings from an explorative investigation that my colleagues and I conducted to understand current challenges and practices around family communication across time zones (Cao et al. 2010)[1]. Based on these findings, we will reflect on the meanings of time difference in connecting families, as well as design opportunities for improving the experience of such cross time zone family communication.

---

[1] Partially reprinted here with permission.

Research Method

We aimed to gain such understandings from people who were already familiar with coping with time differences when communicating with family, therefore, we interviewed 14 people (9 women, 5 men, aged 25–61) who regularly communicate with family members living in other time zones. These were from 12 households, including two households which were related to each other. All participants regularly (ranging from daily to biweekly) communicated with one or more immediate family members (parents, spouse/partner, or children) living in other time zones, with the time difference ranging between (±) 3 ~ 12 h (disregarding date change), large enough to have an impact on communication. For a more holistic understanding, we included participants currently living in four different countries/time zones: UK (Cambridge/London, GMT), US (Seattle, GMT-8), Canada (Toronto, GMT-5), and China (Beijing/Shanghai, GMT+8). The family members they communicated with lived in locations covering nine different time zones in total. Some participants (e.g., grown-up children) had moved from their place of origin and communicated with family back in their original time zone (five households); some (e.g., parents) remained in their native location and communicated with family in other time zones (four households); and for some (e.g., couples) both parties had moved away from their home time zones (three households). Participants' occupations included teacher, researcher, student, IT professional, businessperson, housewife, and retired, resulting in a variety of daily schedules that may influence communication behaviors.

We interviewed the participants either in person or over the telephone. Participants from the same households were interviewed together. The interviews were semi-structured, and each of them lasted about 1 h. Participants were asked to describe their communication experience with each regularly contacted family member in other time zones, such as communication methods, coordination strategies, and so on. Where applicable, they also compared these experiences to communicating with family members living remotely but in the same time zone. All interviews were recorded and transcribed, and then analyzed in order to identify emerging themes using open coding (Strauss and Corbin 1998). Unsurprisingly, these interviews revealed a great deal of more general family communication practices, many of which resonate with the other chapters in this book. In this chapter, however, we focus on extracting experiences directly related to time difference, and report them around the themes that emerged in our analysis.

Findings

Time difference was considered a challenge for family communication by all participants. The main difficulty came from the misalignment of daily schedules between the two parties of communication. Unlike families living in the same time zone whose daily schedule and availability for communication may roughly match, cross time zone families relied on the intersection of their leisure time which was shifted by the time difference. This results in a much smaller and somehow rigid time window available for communication. Our participants have adapted their communication practices to address this challenge, as detailed below.

Reliance on Synchronous Communication

A variety of communication methods were used by our participants to connect with their families, including both synchronous methods such as telephone and internet audio/video call (e.g., Skype™), and asynchronous methods such as email or short message service (SMS). Despite the difficulty posed by time difference, synchronous methods dominated family communication for most participants. This was explained by the nature of family communication, the content of which is mainly emotional contact and catching up about daily life, rather than functional information exchange. Being able to hear the person's voice and to see their face, as well as the real-time interactivity in audio/video conversations proved essential for the sense of presence, connectedness, and dedication between close family members, compared to which the actual conversation content can be secondary. As an extreme example, some couples would leave a live audio or video link on without actually talking to each other, solely for the feeling of presence, as also reported extensively in Greenberg and Neustaedter's chapter on video chat in long-distance relationships. These synchronous calls were treated as a dedicated activity and almost always happened at people's homes. The typical length of a conversation varied from 10 min to about 1 h for different participants. Similarly, instant messaging (IM), which can be seen as the middle ground between synchronous and asynchronous communication, was more often used synchronously in dedicated chat sessions.

By comparison, asynchronous communication was recognized as more flexible because it only required one party to be available, and therefore could be initiated outside the "communication window" dictated by the time difference. However, in practice these were used much less frequently than synchronous communication methods for the reasons mentioned above. Our participants said they would often rather wait to make a call than opting to send an asynchronous message. This was in contrast with cross time zone communication in work settings, where email constitutes a major part of the communication. In family communication, we found that asynchronous communication was mostly used either to make up for a missed or long overdue call ("*If I have been really busy and haven't had time to call them… I drop them an e-mail*"), or to notify about temporary unavailability for a future call.

This domination by synchronous communication can be seen as a testament of family values, which emphasize on emotional connection rather than necessarily conveying information. Indeed, this reliance sets the basis of all other practices we discuss later. Comparatively, although investigations on cross time zone collaboration in work settings (Tang et al. 2011) have also observed a necessity of synchronous meetings, there the rationale is to maximize communication efficiency and avoid delays, a classic exemplification of work values. In the context of family communication, however, efficiency is perhaps the last factor to be considered.

Implicit Coordination through Routines

This preference for synchronous communication requires coordination in finding the time slot to accommodate family members in both time zones. However, different from work settings where people carefully negotiate the time beforehand for international phone calls, we found that, in the family environment, the actual communication time was almost never explicitly negotiated in advance. Instead, our participants relied on implicit "soft routines", where a relatively regular time window was informally recognized by both parties as an appropriate range within which to call, e.g., 10–12 am for one party and 6–8 pm for the other in the case of an 8-hour difference. However, the exact time of the calls was not fixed. The call could be initiated at any time during the "communication window". People tried to make themselves available during the communication window, and would inform the other party in advance if they would not be. Sometimes, IM status was also used to reconfirm availability during the window, especially if the call was going to be made using the computer itself. In most cases (especially inter-generation communication cases), these communication windows were during the weekend since there was a larger range of free time to choose from for both parties, naturally leading to a weekly communication pattern. For families with a large time difference (e.g., >5 h), the intersection of leisure time on workdays was often nonexistent or too short to be feasible. Depending on the time difference and participants' daily schedule, the length of these communication windows varied from 1–2 h to half a day.

Such communication routines gradually emerged over time, but were never explicitly agreed on. For the routine to be established, knowledge about the other party's daily (and weekly) schedule was important. All our participants were able to describe the typical daily schedule of the remote family member at varying levels of detail, and they used this information to facilitate communication. For people communicating back to their original home, this knowledge mostly came from the previous experience of living together ("*It was like that when I lived at home*"). This was less useful for people (e.g., parents) communicating with family members who had moved away, since moving to a new location usually also implies dramatic changes in daily life routines. For them, this knowledge was accumulated over time after the move, both from the communication patterns that emerged, and from casual mentions of daily events during conversations. Some of our participants found it surprising how much detail their parents back home knew about their daily schedule, even though they had never shared it intentionally!

Although communication with family had become an integral part of their lives, our participants considered it secondary to other daily routines. They typically would not change their own schedule in order to accommodate communication with remote family members, except for special occasions such as New Year's Eve.

Similarly, they would not try to contact their family at an inappropriate time for them, especially during hours of sleep, even if there was an urgency to talk.

Of specific interest was when participants' daily schedule changed. When the participants had to temporarily adjust their schedule or plan activities that would impact on their usual time window for communication, they almost always notified their remote family members in advance, either in a previous conversation, or through asynchronous channels such as emailing or instant messaging. In most cases the conversation was cancelled and people would simply wait until the next routine time, since to reschedule the conversation outside the routine window would most likely require explicit negotiation, and people do not perceive the justification of this extra effort. In our study there were also four cases where people had permanent changes in daily schedule when they went through changes in life, such as graduation or retirement: "*I started out as a postgraduate and my time was quite flexible… now it's clear, you know nine to five it's at work, outside of that it's at home.*" In these cases, a new communication routine gradually emerged to adapt to the change, similar to how routines formed when people first moved.

As a special case of schedule change, many participants mentioned travelling as an additional challenge for communication. When one of the two parties was travelling, not only were they likely in an unfamiliar time zone, but also their daily schedule would become much less regular than at home. Combining these two factors, their availability for communication would become completely unpredictable for the other party, and the established communication routine would be entirely broken. As a result, most people opted not to communicate during travel at all, or solely relied on asynchronous channels such as email. Travel also often led to the traveler being called at inappropriate times if the other party was not properly informed.

One might wonder why families opt for such a seemingly inefficient and anarchic way to coordinate the cross time zone communications. In fact, such implicit routines also reflect the nature of family relationships, which build on emotional obligations rather than explicit protocols. On the one hand, the soft routine ensures that the emotional obligations for each other get implemented; on the other hand, the lack of an explicit protocol retains the feeling that the communication is voluntary and indeed driven by emotional needs. In a sense, people want to "be obligated without being obligated".

Being Sensitive to Time

Our participants were all well aware of the exact extent of time difference between them and their family. To convert time between the two time zones, different people developed different mental systems to ease the calculation. For example, for one couple, a 16-hour difference was calculated as "*minus 8 and add another day*" by the husband, and "*day and night switch and another 4 h*" by the wife. Most participants did the conversion in their heads, while a few used digital or paper tools to facilitate the conversion, such as displaying multiple clocks on the computer, or drawing a conversion chart. Experience living in the relevant time zone seemed to greatly help with the ability to do conversion. As a result, people who communicated with their original time zone were generally more effective with the conversion than those (especially parents) who remained in the native time zone and communicated with family members living away. For the latter, having temporarily visited the other site usually also resulted in improvements in the conversion ability. Although time conversion was usually not a big difficulty for regularly communicating family members, it was often a challenge for less-experienced older adults such as grandparents. Several participants recalled being wakened in the middle of the night by phone calls from grandparents, who were then "*too afraid to ring after that*".

It was interesting to hear participants' thoughts about different extents of time difference, especially from those who had experienced more than one (e.g., due to moving from one foreign country to another, or communicating with multiple family members living in different locations). Contrary to intuition, a longer time difference was not necessarily considered worse. A "good" time difference was one that conveniently matched the leisure time of both parties. For example, a 12-hour difference, the longest possible when disregarding date change, was actually considered one of the better cases since it matched up free time in morning and night between the two sites. With the two time zones being exactly symmetric in the day, it also created two communication windows per day instead of one. In addition, the 12-hour difference was one of the easiest to calculate by simply inverting the am and pm. In contrast, an 8-hour difference was considered amongst the worst cases, resulting in either party being working or sleeping at any given time on a regular workday.

When mentioning a particular time to their family members, especially for co-ordinating communications, all our participants referred to it by converting to the other time zone, or repeating the time for both time zones. Only when the event was completely irrelevant to the other party would they refer to it by local time only. During conversations, people often referred to the time as well as associated activities at the remote site ("What time is it?", "Have you had dinner?", "You should go to bed now."). This helped to set the context of the conversation, and was a casual topic of conversation to show their sensitivity and awareness to the other.

As we showed, these various forms of sensitivity to the time on the other side are not only functional, but can also be seen as a way for people to display their consideration and dedication for their remote family members, a sign of "putting myself in your shoes". This again demonstrated family values in the communication practice, values that emphasize on showing care.

Less Regularly Contacted Family Members

Although we focused on immediate family members who communicated heavily with each other, our participants often also mentioned other family members who they communicated with less regularly across time zones, a frequent example being siblings. Especially within the younger generation, siblings usually feel little obligation for dedicated communication with each other, and relied more on ad hoc communication such as through IM. As a result, time difference had less impact on their communication pattern. Instead of having knowledge about each other's schedule, IM status became the main source for them to check availability for conversations, which were then conducted through IM chat or audio/video calls. The actual local time of the other party was usually not taken into account. On the one hand, this was a natural reflection of the less regular life style of many young adults today. On the other hand, interestingly these were the same people who abided by regular communication routines with their parents or significant others. There seems to be a certain level of impression management involved, in that people try to maintain the impression of a healthy and regular life style in front of family members who care about them most, by not calling outside the routine communication window even when they know both parties are indeed available.

Connecting Within the Same Time Zone

For a comparison, our participants also described their experience telecommunicating with remote family members in the same time zone. In contrast to cross time zone communication which is a dedicated activity and has a relatively rigid routine, same time zone family communication tended to be much more flexible and ad hoc. Without the constraint of a small communication window, people had shorter and more frequent communications throughout the day, which happened at home, at work, or in transit. Relatively little beforehand planning was needed to choose the communication time, since people could simply check again at a later time if the other party was not available at that moment. As such, knowledge of the other party's daily schedule played a much lesser role. This resulted in very different communication dynamics, where such lightweight exchanges complemented less frequent intense conversations. The lightweight communication kept people connected about thoughts and feelings on the spur of the moment, or simply for people to express care, which were both critical components of emotional connection. It also improved the experience of the more dedicated communications, by keeping the conversation flow going, and providing more context and topics for the conversation—the more you talk, the more you have to talk about.

Indeed, this type of lightweight communication is also common between family members living together, but almost completely lacking in cross time zone family communication. Despite all the strategies for coping with time differences that families are adept at, this is one key limiting factor that cannot be easily overcome, which may compromise the communication experience and sense of connection.

Meanings of Time Difference

Based on these findings, we can now reflect on some of the deeper meanings of time difference in connecting families. As we will see, time difference is more than a "problem" to solve, but rather give us a lens to look into many of the meaningful aspects of family connection in general.

Time Difference as a Testament of Family Values

As already alluded to when discussing our research findings, family values are a ubiquitous factor in defining the practices around connecting families across time zones. The emphases on connecting emotion rather than exchanging information, fulfilling commitment rather than attending appointments, displaying sensitivity rather than maximizing efficiency, can all testify the prevalence of family values: caring, loving, and supporting; as opposed to work values: accomplishing, optimizing, and negotiating.

This dichotomy in values behind communication resulted in a very distinct pattern when families deal with time differences, often choosing seemingly inefficient or inconvenient solutions such as relying on synchronous rather than asynchronous conversations, or employing soft communication routines without explicit agreement. However, it is precisely through these rituals that people elaborately reconstruct the bonds between family members, which may otherwise be weakened by the separation of time zones.

It is also interesting to consider people who need to cope with time differences in both work and family settings. This is especially common for expats working in satellite teams, who need to regularly communicate both to the headquarters and to their families back home. The same person may employ very different strategies when dealing with the same challenge with colleagues and family, and often need careful planning to reconcile both in a consistent daily schedule. This may pose special design challenges for creating communication tools for this population group.

Time Difference as a Separator and a Connector

Every coin has two sides, as is time difference's influence on connecting families. One of the most meaningful aspects of such connection is connecting the seemingly mundane life of each family member. We already witnessed how time difference severed the continuous lightweight communication between family members, and forced them to rely on a limited number of discrete conversations based on routines. This doubtlessly separates family members whose life would otherwise be more tightly intertwined. However, the very same time difference may also serve as a connector to encourage family members to learn about each other's life. As described previously, in order to establish the communication routines, people need to first build a good understanding of the other party's daily life schedule. It is exactly through this learning process that people maintain the empathy for their remote family members' daily life, which otherwise could become lost in the long-distance separation.

Time Difference as a Player in the Larger Ecosystem

Needless to say, time difference is only one of the many factors that define the long-distance family communication experience. Geographical, social, and cultural contexts all come into play in determining how the remote family members are connected, and they interact with the influence of time difference.

One simple example of the geographical context was that in large countries that span multiple time zones themselves, the experience with time difference may be more familiar to the general public, if not through direct travelling experience then through countrywide TV or radio broadcasting. In contrast, in smaller and more geographically isolated countries, the notion of time difference is only to be experienced when people or their family members travel abroad. This may have a direct impact on people's adeptness with the time difference.

The social and economic context may interplay with the effect of time difference in various ways. One dimension is the familiarity and accessibility of computer-based communication tools. In fact, many people of the older generation started using computers solely for connecting with their children or grandchildren. Due to the lack of proficiency with technology, they were put into a more passive position in determining the communication patterns, often relying on their children to initiate the conversation. Another example is that in some developing societies, facilities for making synchronous audio/video calls are only available in internet cafes but not in many homes, which would obviously have an impact on how families plan their communication across time zones.

The cultural context is yet another important factor to consider. For example, the division between work and life is recognized differently in different cultures: while in some cultures it is perfectly acceptable to call family in work places, in other cultures this is considered a taboo. This would have an influence on people in how they perceive certain time windows as suitable for family communication, and in turn affect the communication routines. In addition, the surrounding culture may also affect the communication patterns by imposing certain daily schedules to the people, such as dinner times.

Design Opportunities

Indeed, since our investigation, some researchers have started designing family communication technologies either specifically to cope with time difference, or with time difference in mind. For example, CU-Later (Tsujita et al. 2010) is a system to allow synchronizing activities across time zones by displaying recorded video of a remote activity after a time shift, such as connecting two remote dining tables and letting family members see and hear each other having dinner despite actually having done so at different times; Family Window (Judge et al. 2010) is a media space that supports always-on live video between two families, while it also allows time shift video recording for users to catch up with activities they missed due to different time zones or schedules; CoupleVIBE (Bales et al. 2011) is a mobile application designed for long-distance couples, which automatically pushes a person's location information to her partner's mobile phone via vibrotactile cues, to compensate for the lack of continuous lightweight emotional connection; and Toaster (Raffle et al. 2011) is a jack-in-the-box toy with an embedded mobile phone to make asynchronous messaging more playful and emotionally meaningful for young children, which incidentally is also one of the user groups that may suffer most from time difference, given their days are usually shorter compared to adults. These are merely a few examples of how family communication tools can be designed to be sensitive to time differences.

In addition to these, inspired by the unsolved challenges we identified through our research, we also list two of the interesting design opportunities that we feel may help improve the current cross time zone family communication experience.

Awareness of Exception to Routines

As we found, people had good knowledge about the typical daily schedule of their remote family members, which was critical for them to establish their communication routine. However, whenever these daily routines were temporarily broken, the extra effort required to renegotiate the conversation time often led to cancellation of the communication. Lightweight methods to help family members be aware of and deal with exceptions could be very valuable. This may take the form of a precaution message to remind the other party of schedule change in advance, or as a just-in-time warning to prevent calling at inappropriate times. For example, travelers might benefit from a mobile phone that leveraged location data to warn callers of the local time during late night hours. For example, "It's 22:00 for Susan right now, do you want to complete this call or leave a message". More generally, communication tools might provide more assistance visualizing the alignment of typical daily schedules to identify otherwise overlooked alternative communication times.

Lightweight but Timely Communication

Ad hoc lightweight communication appears to have an important role in same time zone family communication, not only to keep each other updated but also to demonstrate caring. It is interesting to speculate how we might enable similar kinds of communication for cross time zone situation as well, e.g., by sending short video or voice messages. However, the content of such lightweight communication is often trivial and only meaningful when put in the current temporal context, a possible reason why such communication was not common in cross time zone situations. We could consider an asynchronous messaging service that delays the delivery so that the message arrives at a suitable time for the receiver. For example, a person could send her spouse a morning greeting voice message in her own morning, but the message would only be delivered when it becomes morning in the other time zone. Another possibility is to accumulate numerous lightweight messages over a day or a week, and deliver them as a collection periodically, so that subtle feelings to be communicated never "miss the moment" when they emerge.

Obviously, these two design opportunities should not be seen as exclusive or prescriptive, but merely examples of the rich design space.

Conclusion

Despite the fast advances in communication technology, time difference remains one of the few challenges in telecommunication that will likely never be truly "solved". On the contrary, its influences will only become more prominent as more and more families have ready access to modern communication technologies. Therefore, understanding the role of time difference in connecting families can be regarded as both a timeless and a timely thesis, to which this chapter aims to bring more attention.

References

Bales, E., Li, K. A., & Griwsold, W. (2011). CoupleVIBE: mobile implicit communication to improve awareness for (long-distance) couples. *Proceedings of the ACM conference on Computer supported cooperative work (CSCW)* (pp. 65–74). New York: ACM. doi:10.1145/1958824.1958835.

Cao, X., Sellen, A., Brush, A. J. B., Kirk, D., Edge, D., & Ding, X. (2010). Understanding family communication across time zones. *Proceedings of the ACM Conference on Computer Supported Cooperative Work (CSCW)* (pp. 155–158). New York: ACM. doi:10.1145/1718918.1718947.

Judge, T. K., Neustaedter, C., & Kurtz, A. (2010). The family window: the design and evaluation of a domestic media space. *Proceedings of the international conference on Human factors in computing systems (CHI)* (pp. 2361–2370). New York: ACM. doi:10.1145/1753326.1753682.

Kim, S., Kientz, J. A., Patel, S. N., & Abowd, G.D. (2008). Are you sleeping?: sharing portrayed sleeping status within a social network. *Proceedings of the ACM conference on Computer supported cooperative work (CSCW)* (pp. 619–628). New York: ACM. doi:10.1145/1460563.1460660.

Lindley, S. E., Harper, R., & Sellen, A. (2009). Desiring to be in touch in a changing communications landscape: attitudes of older adults. *Proceedings of the international conference on Human factors in computing systems (CHI)* (pp. 1693–1702). New York: ACM. doi:10.1145/1518701.1518962.

Lottridge, D., Masson, N., & Mackay, W. (2009). Sharing empty moments: design for remote couples. *Proceedings of the international conference on Human factors in computing systems (CHI)* (pp. 2329–2338). New York: ACM. doi:10.1145/1518701.1519058.

Modlitba, P. L., & Schmandt, C. (2008). Globetoddler: designing for remote interaction between preschoolers and their traveling parents. *Extended abstracts on Human factors in computing systems (CHI EA)* (pp. 3057–3062). New York: ACM. doi:10.1145/1358628.1358807.

Raffle, H., Ballagas, R., Revelle, G., Mori, K., Horii, H., Paretti, C., & Spasojevic, M. (2011). Pop goes the cell phone: asynchronous messaging for preschoolers. *Proceedings of the ACM International Conference on Interaction Design and Children (IDC)* (pp. 99–108). New York: ACM. doi:10.1145/1999030.1999042.

Strauss, A. C., & Corbin, J. (1998). *Basics of qualitative research*. Thousand Oaks: Sage.

Tang, J. C., Zhao, C., Cao, X., & Inkpen, K. (2011). Your time zone or mine? A study of globally time zone-shifted collaboration. *Proceedings of the ACM Conference on Computer Supported Cooperative Work (CSCW)* (pp. 235–244). New York: ACM. doi:10.1145/1958824.1958860.

Tsujita, H., Yarosh, S., & Abowd, G. D. (2010). CU-Later: a communication system considering time difference. *Proceedings of the ACM international conference adjunct papers on Ubiquitous computing (Ubicomp)* (pp. 435–436). New York: ACM. doi:10.1145/1864431.1864474.

Zerubavel, E. (1985). *Hidden rhythms: schedules and calendars in social life*. Berkeley: University of California Press.

Chapter 8
Inter-Family Messaging with Domestic Media Spaces

Tejinder K. Judge, Carman Neustaedter and Steve Harrison

Abstract Many family members have a need to stay connected with their loved ones when they are separated by distance. Technologies such as the phone or email help achieve this to some extent, but, many people still feel out of touch with their loved ones. We designed two domestic media spaces—The Family Window and Family Portals—to help distributed family members connect with remote families' homes using 'always-on' video connections. In addition to this, both systems allowed family members to interact using handwritten messaging. Our chapter focuses on this latter functionality to explore the ways in which family members made use of the inter-family messaging features found within our domestic media space systems. Here we discuss both synchronous and asynchronous messaging and the nuances of public vs. private messaging between households. We conclude with a discussion of implications for inter-family messaging systems.

T. K. Judge (✉)
Google Inc.,
Mountain View, CA, USA
e-mail: tkjudge@google.com

C. Neustaedter
School of Interactive Arts and Technology,
Simon Fraser University,
102 Avenue 250-13450, V3T 0A3, Surrey, BC, Canada
e-mail: carman_neustaedter@sfu.ca

S. Harrison
Department of Computer Science and School of Visual Arts,
Center for Human-Computer Interaction,
Virginia Polytechnic Institute and State University,
2202 Kraft Drive, Blacksburg, VA 24060, USA
e-mail: srh@vt.edu

C. Neustaedter et al. (eds.), *Connecting Families,*
DOI 10.1007/978-1-4471-4192-1_8, © Springer-Verlag London 2013

Introduction

Many families and loved ones who are separated by distance try to remain connected and aware of each others' lives in order to feel closer to one another. This includes sharing and learning about one's activities, locations, and status (e.g., health) (Neustaedter et al. 2006; Romero et al. 2007; Tee et al. 2009). For example, parents may want to know about the well-being of their adult children who have 'left home' to live independently or start their own families. Similarly, grandparents often want to learn about their grandchildren as they grow up and know what type of extra curricular activities they are participating in, how their schooling is going, etc. This is elaborated on in Moffatt, David, and Baecker's chapter on connecting grandparents and grandchildren. In addition to the sharing of information, people also typically still want to participate in family gatherings such as holiday get-togethers, birthday parties, and other social gatherings. However, such family gatherings are easily missed unless one is able to travel.

Families use a variety of technologies to stay connected with their loved ones over distance. The phone allows family members to synchronously communicate and discuss each other's lives and happenings. Email supports the asynchronous sharing of information. Instant messaging affords both synchronous and asynchronous communication depending on how family members utilize the technology. While all are beneficial technologies, none allow family members to actually see each other, akin to the way they might in face-to-face situations. The act of being able to see another family member has been shown to provide additional feelings of closeness (Neustaedter et al. 2006; Tee et al. 2009; Ames et al. 2010; Judge and Neustaedter 2010; Kirk et al. 2010).

It is for this reason that many families have begun to adopt off-the-shelf video conferencing, or 'video chat' systems, such as Skype, Apple FaceTime, and Google Chat, to stay connected with their remote family members. Yet the challenge is that most are designed to be used in a manner similar to the telephone where one calls another person for a fixed time period. Such design and implied usage makes video calls limited when it comes to sharing longer activities and time periods with remote family members. For these reasons, our research has explored the design of video chat systems where the video link can be easily left on for an extended period of time, akin to media spaces originally designed for the workplace in the 1980s and 1990s (Harrison 2009). We call these domestic media spaces.

First, we designed a *dyadic* domestic media space called THE FAMILY WINDOW that provided an always-on video connection between *two* households using a situated display (Judge et al. 2010). The Family Window also provided a messaging feature where families could leave messages for each other by handwriting on top of the video display. Our field deployment showed that families enjoyed being able to see their remote family members on a daily basis and the messaging feature allowed them to share additional information including greetings, comments, and heartfelt messages. Second, and building on this research, we designed a *multi*-family media space called FAMILY PORTALS that provided a video link between *three* families' homes in addition to *both* private and public messaging capabilities (Judge et al. 2011). Again, our field deployment showed success, though with an increased set of relationships being supported by the system, additional privacy concerns arose. We also saw families adopt distinct messaging practices in terms of when they chose to send messages to each household in a private fashion and when they would publicly send messages to both.

Our focus in this chapter is on describing the ways in which family members adopted and used the messaging features found in both the Family Window and Family Portals as it relates to asynchronous usage, synchronous usage, and private vs. public messaging. For more results on the ways in which family members used the video connection within these systems, we refer readers to our conference papers on the topic (Judge et al. 2010, 2011). We begin the chapter by describing related on work on intra and inter-family messaging, which compliments Schatorje and Markopoulos's earlier chapter in this book on Family Circles. Second, we outline the Family Window's design and our findings on the ways in which families adopted and appropriated its messaging capabilities. This highlights the value of providing messaging capabilities within an awareness system focused around a video connection. Next, we outline the design of Family Portals where we describe the effects of a having a triad of families use the media space for inter-family messaging and the public and private nature of messages. We conclude the chapter by discussing the implications of these practices for the design of future inter-family messaging systems.

Related Work

First, several research prototypes have been designed to support situated *intra*-family messaging. That is, messaging between family members who live in the same residence. TxtBoard was a messaging system that allowed family members to send messages via the short messaging service (SMS) between a situated display in the home and family members' mobile phones (O'Hara et al. 2005). In a field trial, family members used the system to share messages about their location, activity, and status. Following this, Sellen et al. (2006) created HomeNote, which built on TxtBoard's messaging capabilities and added the ability to leave handwritten messages on the home display. Here field deployments with families found the system was used extensively for sharing awareness information, providing social 'touches' for others, and storing information, amongst a variety of other uses. Overall, the usefulness of HomeNote depended on the family and their specific needs. StickySpots was similar in design to HomeNote, but focused on the importance of location when it comes to the placement of messages within the home (Elliot et al. 2007). With StickySpots, family members could leave messages on any number of interconnected displays placed throughout the home, where the placement of a message would provide additional meaning for it. For example, messages meant for parents could be placed on a display situated in a location that they usually looked at when arriving home from work. Similarly, messages meant for children in a family could be placed on displays near their rooms. More recently, we have seen research that moves away from the 'written messaging' paradigm of the above systems. Family Circles allows family members to record audio messages on round messaging tokens, which can then be placed in locations throughout the home for playback (Schatorje and Markopoulos's chapter). This, again, allows contextual information to be associated with the messages. We refer readers to Schatorje and Markopoulos's earlier chapter in this book to learn more about the system and its design.

Several research prototypes have also been designed to support situated *inter-family* messaging between homes. Here we are referring to messaging between one or more households where there may be more than one distinct family unit involved. This is akin to the way that the Family Window supports family messaging. The earliest system, CommuteBoard (Hindus et al. 2001), provided a shared whiteboard for connecting two households. This system allowed carpoolers to leave handwritten messages for one another to coordinate rides. Deployments found that the use of colored digital ink and the informal nature of handwritten notes caused a form of playfulness to appear. However, the legibility of handwriting and limited writing spaces caused usability issues for family members. In their evaluation of SPARCs, a photo and calendar-sharing prototype, Tee et al. (2009) also deployed MessyBoard (Fass et al. 2001) as a comparison to SPARCs. While not originally designed for families, for this evaluation, MessyBoard provided families with a shared messaging board that allowed typed notes to be left for remote family members. The field deployment found that people enjoyed being able to asynchronously leave messages for the remote household (Tee et al. 2009).

While the above systems supported dyadic family connections, the related research also provides examples of inter-family messaging systems that connect multiple households together. Here we are referring to messaging between more than two families, akin to the way that Family Portals supports family messaging. First, messageProbe (Hutchinson et al. 2003) allowed multiple families to leave hand-written messages on "Post-It" notes placed on a canvas shared by all households using the system. In this way, families could see *all* messages posted to the system, but there was no means to send private messages intended for only one household. Second, Wayve (Lindley et al. 2010) allowed families to leave handwritten notes for one another on interconnected messaging appliances, one in each household. Messages could also be sent from Wayve to email accounts or mobile phones and vice versa. In this way, sending from the device could be private if directed to one person's email or phone. Yet all email and text messages sent to the messaging appliance were inherently public to all families. Thus, there was no way to send a private message to a family's situated display. When evaluating Wayve, Lindley et al. (2010) found that most messages were public messages sent between families' Wayve devices, with fewer messages sent privately to individuals via email/phones. What remains unknown is whether or not such behavior would stay consistent if the situated messaging appliances could receive private messages.

Our work builds on the existing research in two ways that form the focus for remainder of the chapter. First, we explore the usage of inter-family messaging systems that are coupled with video media spaces by looking at the design and field trials of both the Family Window and Family Portals. Second, we directly explore families' behaviors when they have the ability to send both private and public messages to displays situated in other families' homes as was possible with Family Portals.

Situated Messaging in a Dyadic Media Space

The Family Window was designed to be a *dyadic* media space that connected two homes with always-on video. Figure 8.1 shows the system being used by a set of grandparents, their children, and grandchildren. The video link from the grandparents' home (the remote view) is shown spanning the majority of the display and a feedback view of the children/grandchildren's home (the local view) is shown in the bottom left corner of the screen (Fig. 8.1). The system runs on a dedicated display such as a tablet PC or digital frame in order to act as an information appliance, as shown in Fig. 8.2. In addition to the video capabilities, families are also able to leave handwritten messages for each by writing on top of the video display using either a stylus or finger. For example, the red handwriting in Fig. 8.1 is a message written at the children/grandchildren's house and the yellow handwriting is the response written at the grandparent's home. Family members can pick and choose ink colors as well as erase content. These writing capabilities build on ideas from workplace media spaces (e.g., Tang and Minneman 1990, 1991). A video of the Family Window and its interaction can be found in Neustaedter et al. (2010).

In order to understand how families would adopt and use the Family Window, we conducted a set of field trials with three family pairs. Two pairs used the system for a period of 5 weeks and one pair used it for 8 months as a part of its autobiographical design (Judge et al. 2010):

Sister Families The first pair connected the families of two sisters, which included connecting two parents and their 18-month-old son with the wife's sister and her long-term male companion. The two households lived a 2-hour drive apart.

Daughter-Parents-Grandchildren Families The second pair connected the families of a daughter and her mother, which included connecting the daughter, her husband, and their 2-year-old son with the child's grandparents. The two households lived in the same time zone, but were a 21-hour drive apart.

Son-Parents-Grandchildren Families The third pair connected the families of a son and his parents. This included connecting the son's wife and their two children, aged 3 years and 8 months (at study completion), with the children's grandparents. The two households were separated by three time zones across North America (coast-to-coast).

We conducted semi-structured contextual interviews with the families throughout their usage and also sent emails and phoned between interviews to ensure

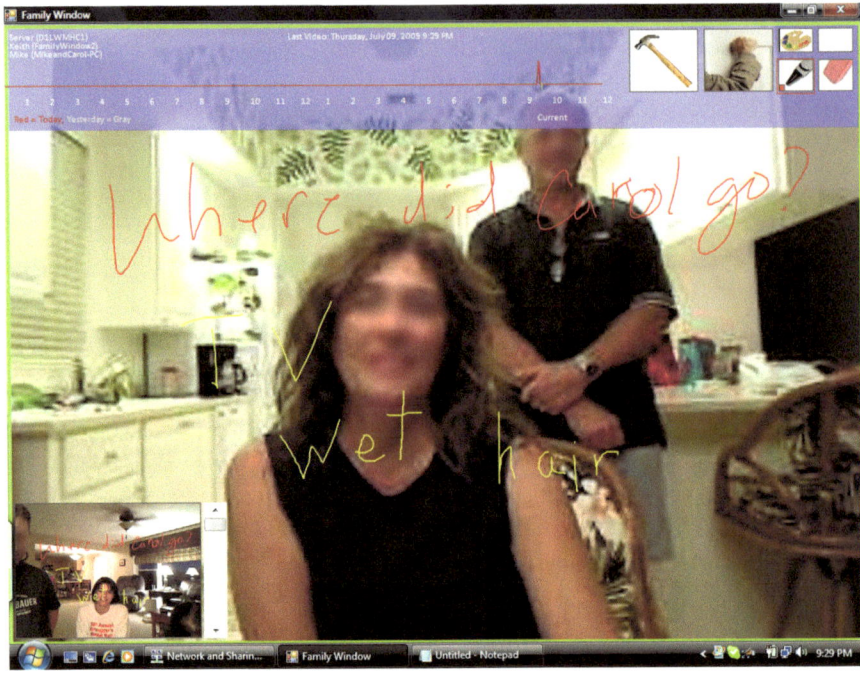

Fig. 8.1 The Family Window's user interface

Fig. 8.2 The Family Window running in a dedicated display in a family's living room

families were not having technical difficulties. We used open, axial, and selective coding to analyze our data and generated codes that reflected a variety of usage patterns (Strauss and Corbin 1998).

Interacting Through the Family Window

The always-on video link in the Family Window provided the families with many opportunities to see what was happening in the remote families' homes, which made them feel closer as a result. Families also used the Family Window as a communication tool for interacting with their remote family members.

First, families often coupled their use of the Family Window with the phone as our design did not provide an audio link (because of assumed privacy risks associated with long-term audio connections). The Family Window would provide the video link to see family members, gesture, or show items of interest and the phone supported the voice conversation. While beneficial, phone calls only sufficed for situations where family members wanted to have longer conversations. In situations where they wanted to simply say a quick 'hi', they relied on the messaging capabilities of the Family Window. In these situations, phoning the other home would have suggested the need for a longer conversation than was necessary. Thus, the Family Window provided family members with a unique opportunity to still exchange information but not be committed to a long conversation. Here we saw families leave a large number of handwritten messages as a form of asynchronous communication. Messages often began with a simple 'good morning' at the beginning of the day and then evolved into more detailed discussions with messages left at various points in reply to one another. Participants told us that seeing these messages in the context of the remote family's video made them special because it was a dedicated communication portal with the remote family. Families also said that these messages required less effort to write than their normal exchanges of email.

> It is nice to come home or wake up to see a message from [my sister]. A simple message like 'have a nice day' is all I need to know that she is thinking of me.—Sister 1 in the Sisters Pair

We also saw instances of synchronous communication occur where families would leave a series of messages one after another in a turn-taking fashion over a series of several minutes. In essence, they had turned the Family Window's messaging canvas into a handwritten 'chat window.' Such chat sessions often progressed slowly (handwriting is often slow), though family members commented that despite the lack of speed, being able to see the remote family member's handwriting presented enhanced feelings of closeness.

In several instances, we learned that the Family Window's messaging capabilities led to an interesting routine for the 2-year-old grandson and his grandmother in the second family pair. The grandson would have exchanges with his grandmother where she would write alphabet letters on the Family Window for him, draw shapes, or hold up different colors to try to teach him new things. In turn, he would draw pictures for her. This routine became so important to the grandson that he would run to the Family Window each day after returning home from daycare, scribble a message on it, and kiss the video of his grandmother's face. If his grandmother was not around, his father would call her house and tell her that her grandson was looking for her. This further illustrates the value that families found in having messaging coupled with the video link.

Situated Messaging in a Multifamily Media Space

Following from our Family Window research, we wanted to understand how media spaces and family messaging would extend beyond a dyad to connect multiple families. We knew from prior research that people like to stay aware of the lives of their remote family members, however, it is not the case that people share the same information with all of their remote family members (Neustaedter et al. 2006; Tee et al. 2009). Different people receive different information and at different frequencies (Neustaedter et al. 2006). For example, an adult child might talk with her mother on a daily basis on the phone, telling her about major happenings each day. On the other hand, the same person might only talk with her grandmother once a month. The information shared in this case will likely be more superficial and focus on specific things like how her children are doing and activities they are involved in at school (Neustaedter et al. 2006). What is unclear is how these findings extend to the use of new situated messaging systems for families. That is, if family members are able to send different information to different families, will they do so and in what ways?

As a first step to answering this question, we designed a new media space called Family Portals that built on the Family Window's design to connect three households together instead of just two (Judge et al. 2011). One could imagine extending this design further to support n-connections, though such extensions are certainly non-trivial (e.g., networking complexities, visualization challenges, privacy issues). Figure 8.3 shows the user interface for Family Portals, which again ran in a tablet display to prototype the idea of a dedicated information appliance. The system provides always-on video feeds between three families' homes, in addition to both public and private messaging features.

Private Messaging The left side of the screen in Fig. 8.3 shows two Targeted Portals (top and bottom), one for each family that a local family is connecting to. The portals show the video feed from the remote home and local family members can leave handwritten messages for specific families by writing on top of their video feed using either a stylus or finger. Only the target family sees the writing; thus, it is a private writing space for the two families. A notification appears at the bottom of the display when a new message is written. Users can pick ink colors and erase writing using the icons on the left side of the Portal.

Public Messaging The right side of Fig. 8.3 shows a Shared Portal. Family members can leave handwritten messages here, which show up for *all* families. Thus,

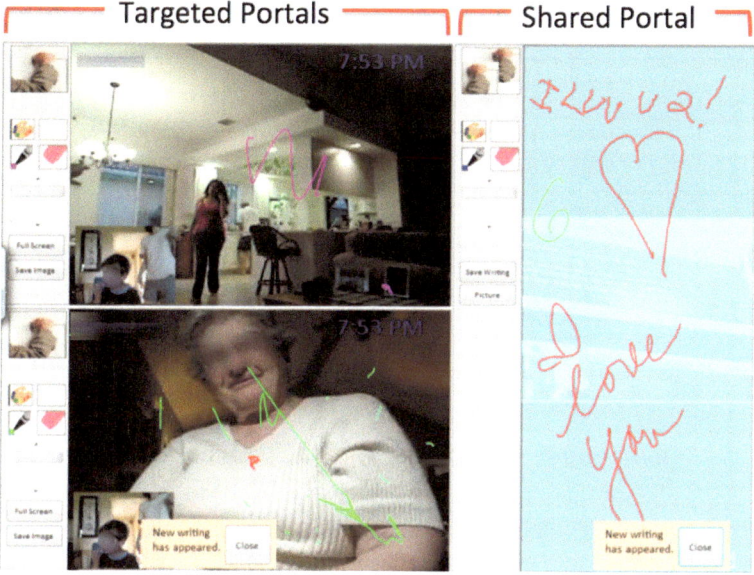

Fig. 8.3 The user interface for Family Portals

it is a public messaging board and offers the same basic functionality as message-Probe (Hutchinson et al. 2003) and Wayve (Lindley et al. 2010). Ink options can be selected to the left of the Shared Portal along with the ability to choose a background picture, which is seen by all families. This picture is overlaid with a semi-transparent blue rectangle so writing is more easily visible.

We conducted a field study with Family Portals in order to learn how families would use its video and messaging features. We recruited six families—two triads—from the USA where all six families used Family Portals within their homes over a period of 8 weeks, though technical issues caused the system to not work during the first 2 weeks. Family compositions are shown in Table 8.1 along with their locations and the pseudonyms we use to refer to the families throughout our results. Triad 1 was composed of women from three generations (daughter, mother, grandmother) and their family members. Triad 2 was composed of the families of two sisters and their mother. All families contained children of varying ages along with partners.

We again conducted semi-structured contextual interviews with the families throughout the course of the field trials. Usage of features was logged and screenshots of writing on Family Portals were also captured by the system. We used open, axial, and selective coding to analyze the interviews (Strauss and Corbin 1998). Next we describe our study results related to family messaging.

Table 8.1 Field study families for Family Portals

	Family name	Household composition	Location
Triad 1	Daughter family	2 parents in 30s, 1 son aged 3	City1, New York
	Daughter parents family	2 parents in 50s	City2, New York
	Daughter grandparents family	2 grandparents in 80s	City3, Florida
Triad 2	Younger sister family	2 parents in 30s, 1 son aged 3	City1, New York
	Sister mother	1 parent in 50s	City4, New York
	Older sister family	2 parents in 30s, son aged 10, son aged 6, daughter aged 1	City4, New York

Public Asynchronous Messaging in the Public Space

The basic usage of the Shared Portal or shared whiteboard was to write messages, questions and notes intended for *all* families. We found this pattern of use among families in both triads. This is similar to messaging practices found with message-Probe (Hutchinson et al. 2003) and the Family Window (Judge et al. 2010). The most common messages were greetings between families such as 'good morning' or 'good night.' Figure 8.4 shows a goodnight message left by the wife in the Daughter Parents family for both the families she was connected to.

Families also used the Shared Portal to share information about where family members were going and what they were doing that day. For example, the husband in the Daughter family wrote one evening,

[Wife] + [son] should be home at 5:30. I'm leaving to teach tonight ☺.—Message written on the shared whiteboard by Husband in Daughter family

Another common use of the Shared Portal was for families to share information about food they were having for dinner and playfully compared each other's menus. For example the wife from the Daughter family wrote one night, *"What's for dinner? Ckn nug [chicken nuggets] & tater tots here…"* and her parents responded, *"M&D [Mum and Dad] having wine."*

During the first few weeks of usage, families faced some confusion over the author of messages on the Shared Portal. For example, it was difficult for families to determine the author of a message if it was written in all capital letters or if the content of the message was general to all families. Some family members left their initials at the end of a message, but over time, this became unnecessary as families learned to recognize each other's handwriting or used the context of the message and their shared common ground to determine the author.

Fig. 8.4 Good night greeting from wife in Daughter Parents family

Private Asynchronous Messaging in the Public Space

We also found that families used the Shared Portal for messages intended for a *specific* family, even though the third family could see them. This pattern of use was mainly found in Triad 2. For instance, Sister Mother lived close to Older Sister's family and met them 3–4 times a week. She and Older Sister would use the Shared Portal to schedule their meetings. They did so without worrying about Younger Sister feeling excluded because Younger Sister knew that her mother frequently visited her sister's family. Figure 8.5 shows one such message written by Sister Mother for Older Sister's family about meeting them at 6 pm one night.

In such cases, families reported that they preferred to write on the Shared Portal as opposed to the Targeted Portal, as they felt messages on the Targeted Portal may be hard to read due to being on top of the video. This suggests a usability issue in terms of readability when multiple information sources (e.g., video *and* writing) use the same region of the display. Yet families also said that in these situations, they did not mind that the third family could see the message on the Shared Portal.

Fig. 8.5 Message to Older
Sister's family from Sister
Mother

Synchronous Messaging in the Public Space

Although we expected that the writing features of Family Portals would mostly be used for asynchronous messaging, we found that families used Family Portals for synchronous interaction akin to 'chat sessions.' This was similar to the use of the Family Window only the chat sessions with Family Portals occurred over larger time spans (e.g., 20–30 min). We believe this difference was idiosyncratic to our participants as opposed to an effect of the difference in systems. Figure 8.6 shows an example from Triad 1 where the wife in the Daughter Parents family is chatting with her mother using the Shared Portal. Most chats were between just two households because it wasn't often that members from all three families would be serendipitously present in front of their Family Portals at the same moment.

Interestingly, families used the Shared Portal and not the Targeted Portal for these dyadic communication episodes. Again, they found it easier to read messages not written on top of the video, but they also said that they were typically chatting about general topics such as family activities, an update after a doctor's visit, etc. In these situations, families were also not concerned about the third family 'walking in' and reading their chats. They told us that if a member from the third family became available at a certain point, they could easily join the conversation by

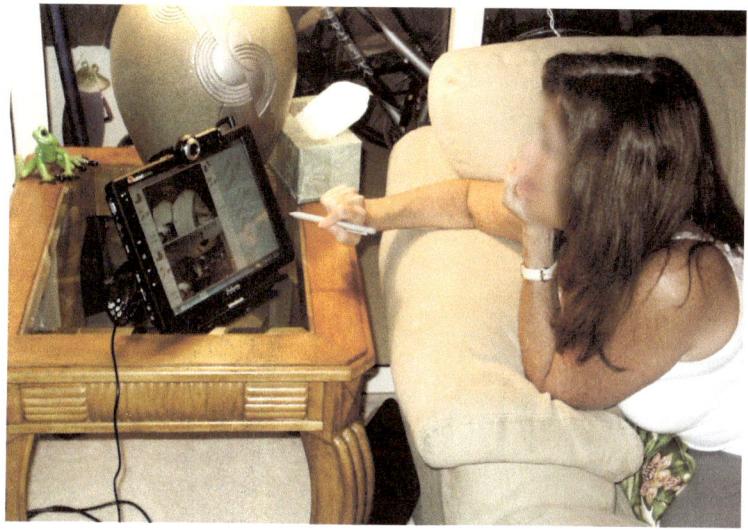

Fig. 8.6 Wife in the Daughter Parents family engaged in synchronous messaging or "chatting" with her mother

reading the previous messages. Families also preferred using the Shared Portal for chats because it allowed them to see each other while writing. Being able to see each other augmented the experience and they did not want to lose this by writing on each other's video feed.

If all three families were present for a synchronous chat, they naturally used the Shared Portal. Participants did not tell us about any situations where Targeted Portals were used as backchannels between only two families when all three were conversing in the Shared Portal.

Confidential Messaging in the Private Space

As one might expect, families did use the Targeted Portals for private messages and discussions that they did not want the third family to know about. For instance, Older Sister and Younger Sister used the Targeted Portal to discuss their suspicion that their mother was not following the diet her doctor recommended. In this case, both sisters would be mortified if their mother would have accidentally seen this discussion.

It was easy for family members to decide where such messages should go given the nature of the information. The readability difficulties of writing on top of the video feed were much less of a concern than the confidential information contained in the messages. In some ways, readability challenges provided a psychological 'cloak,' which visually suggested that the messages were private due to their (sometimes lack of) legibility.

Fig. 8.7 Message on Targeted Portal from wife in Daughter family to her grandmother

Selective Messaging in the Private Space

Families also used the Targeted Portal for situations where they wanted to leave a message for one family, but knew it did *not* involve the third family. They did this to simplify communication and to ensure that the family the message was intended for would easily know there was a message for them. For example, the wife in the Daughter family wrote the note shown in Fig. 8.7 on her grandmother's Targeted Portal. While there was nothing confidential in this message, it was written on the Targeted Portal because it did not involve the third family and was intended *specifically* for the grandmother. Thus, families recognized this and, whether they realized it or not, reduced 'information clutter' for other families.

Families also used the Targeted Portal for topics they had in common with one household and not the other. The shared common ground between the two households made it easy to send these messages and not feel badly about leaving out the third household.

> When I have a question for [daughter] it is easier to write it in her window [Targeted Portal] instead of writing it on the chalkboard [Shared Portal] and having to explain it to my mother.—Interview with wife in Daughter Parents family

Both of these cases were found despite the fact that messages written in the Targeted Portals may be harder to read on top of the video feed. This pattern of use was mainly found in Triad 1. Both the Daughter family and Daughter Parents family made a conscious effort to reduce information clutter for the grandparents to prevent any confusion that might results in them shying away from the technology. Thus, the need to reduce information clutter for families not involved in a conversation superseded the usability issue of writing on top of the video feed.

Discussion and Conclusions

In this chapter, we have explored inter-family messaging when it is coupled with an always-on video link provided by a media space. This has included discussions of the design and evaluation of two such systems, The Family Window and Family Portals. In both cases, we found that families leveraged the messaging features of the system in order to support both synchronous and asynchronous communication. This also revealed the need for families to exchange short messages without being committed to long conversations (e.g., on the phone). Because messages were placed in the context of the remote family—their video link—they had additional meaning and were uniquely associated with that family. In addition to this, we also found challenges with both systems in terms of how they presented their messaging capabilities.

First, families sometimes faced challenges in identifying who was writing messages. This was particularly problematic with the Family Portals because there were multiple households, and multiple family members within them, that might be using the system. Although families were able to resolve this issue over time by learning each other's handwriting and using the context of the message to determine the author, this problem will be more prominent in multiparty messaging system connecting more than three families. This suggests mechanisms that allow families to identify which family members and/or households left which messages. For example, systems could identify different families with different colors.

Second, readability was an important factor for families when choosing where to leave messages. Written messages on top of the video link could sometimes cause readability issues, but this depended on what was being shown in the video feed. This affected where family members wanted to leave messages on the display. Writing on top of the video feed also prevented family members from seeing each other while chatting. Although families are typically not able to see each other while chatting using other tools (e.g., instant messaging), the option to *see* the other person while communicating with them was greatly valued by families.

Third, confidentiality and reducing information clutter were also factors that families considered when choosing where to leave messages. This was seen with Family Portals because of the introduction of a third family. Although writing on the video feed in the Targeted Portal caused readability issues, at times families' needs to send confidential messages superseded this issue. Similarly, the need to selectively target content at one family and not both to reduce information clutter and due to shared common ground, was also more important than readability issues.

Fourth, and more generally, it is clear that families find value in the inclusion of *both* public and private messaging within a family messaging system. This is evidenced by the examples from the Family Portals study and also the fact that family members recognized that even though some content might be directed at one family, it could also be interesting for another family to see. In these situations, families chose a public space for writing, despite the targeted nature of the message. This illustrates that families *are* thinking about who would likely want to see their messages beyond the intended recipient using their judgments to decide where to place messages.

Lastly, our work is certainly not without its limitations. Both systems were used by only a small number of families. While typical for domestic field trials because of their complexity, this does not allow us to more broadly understand how different family compositions and relationships will make use of family messaging systems. Despite this, it is likely the case that families will still value the ability to send both private and public messages, and will continue to value the linkage between video connections showing the remote family and messaging features; however, the specific usage of these features may differ with additional families.

Acknowledgements This research was graciously funded by Eastman Kodak Company. We are also very thankful for the help and support of researchers and management at Kodak Research Labs: Andrew Kurtz, Andrew Blose, Elena Fedorovskaya, and Rodney Miller. Lastly, we are indebted to the families who participated in our field deployments and spent many hours meeting and interacting with us. Without them, the research would not have been possible.

References

Ames, M.G., Go, J., Kaye, J.J., Spasojevic, M. (2010). Making love in the network closet: the benefits and work of family videochat. *ACM Conference on Computer Supported Cooperative Work 2010* (pp. 145–154). New York: ACM.

Elliot, K., Neustaedter, C., Greenberg, S. (2007). StickySpots: using location to embed technology in the social practices of the home. *ACM Conference on Tangible, Embodied and Embedded Interaction*. New York: ACM.

Fass, A., Forlizzi, J., Pausch, R. (2001). MessyDesk and MessyBoard: two designs inspired by the goal of improving human memory. *ACM Conference on Designing Interactive Systems 2001* (pp. 303–311). New York: ACM.

Harrison, S. (2009). *Media Space: 20+ Years of Mediated Life*. London: Springer.

Hindus, D., Mainwaring, S., Leduc, N., Hagstr, A., Bayley, O. (2001). Casablanca: designing social communication devices for the home. *ACM SIGCHI Conference on Human Factors in Computing Systems 2001* (pp. 325–332). New York: ACM.

Hutchinson, H., Mackay, W., Westerlund, B., Bederson, B., Druin, A., Plaisant, C., Beaudouin-Lafon, M., Conversy, S., Evans, H., Hansen, H., Roussel, N., Eiderback, B. (2003). Technology probes: inspiring design for and with families. *ACM SIGCHI Conference on Human Factors in Computing Systems 2001* (pp. 17–24).

Judge, T. K., & Neustaedter, C. (2010). Sharing conversation and sharing life: video conferencing in the home. *ACM SIGCHI Conference on Human Factors in Computing Systems 2010* (pp. 655–658). New York: ACM.

Judge, T. K., Neustaedter, C., Kurtz, A. (2010). The family window: the design and evaluation of a domestic media space. *ACM SIGCHI Conference on Human Factors in Computing Systems 2010* (pp. 2361–2370). New York: ACM.

Judge, T. K., Neustaedter, C., Harrison, S., & Blose, A. (2011). Family portals: Connecting families through a multifamily media space. *ACM SIGCHI Conference on Human Factors in Computing Systems 2011* (pp. 1205–1214). New York: ACM.

Kirk, D.S., Sellen, A., Cao, X. (2010). Home video communication: mediating 'closeness'. *ACM Conference on Computer Supported Cooperative Work 2010* (pp. 135–144). New York: ACM.

Lindley, S., Harper, R., Sellen, A. (2010). Designing a technological playground: a field study of the emergence of play in household messaging. *ACM SIGCHI Conference on Human Factors in Computing Systems 2010* (pp. 2351–2360). New York: ACM.

Neustaedter, C., Elliot, K., Greenberg, S. (2006). Interpersonal awareness in the domestic realm. *Australia Conference on Computer-Human Interaction (OzChi) 2006* (pp. 15–22). New York: ACM.

Neustaedter, C., Judge, T., Kurtz, A., Fedorovskaya, E. (2010). The family window: connecting families over distance with a domestic media space. *Video Proceedings of the Conference on Computer Supported Cooperative Work (CSCW 2010)*. New York: ACM.

O'Hara, K., Harper, R., Unger, A., Wilkes, J., Sharpe, B., Jansen, M. (2005). TxtBoard: from text-to-person to text-to-home. *Proceedings of the Conference on Computer-Human Interaction (CHI 2005), Extended Abstracts*. New York: ACM.

Romero, N., Markopoulos, P., Baren, J., Ruyter, B., Ijsselsteijn, W., Farshchian, B. (2007). Connecting the family with awareness systems. *Personal Ubiquitous Computing, 11*(3), 299–312.

Sellen, A., Harper, R., Eardley, R., Izadi, S, Regan, T., Taylor, A., Wood, K. (2006). Situated messaging for the home. *ACM Conference on Computer Supported Cooperative Work (CSCW 2006)*. New York: ACM.

Strauss, A., & Corbin, J. (1998). *Basics of qualitative research* (2nd ed.). Thousand Oaks: Sage.

Tang, J., & Minneman, S. (1990). VideoDraw: a video interface for collaborative drawing. *ACM SIGCHI Conference on Human Factors in Computing Systems 1990* (pp. 313–320). New York: ACM.

Tang, J., & Minneman, S. (1991). VideoWhiteboard: video shadows to support remote collaboration. *ACM SIGCHI Conference on Human Factors in Computing Systems 1991* (pp. 315–322). New York: ACM.

Tee, K., Brush, A.J., Inkpen, K. (2009). Exploring communication and sharing between extended families. *International Journal of Human-Computer Studies, 67*(2), 128–138.

Chapter 9
Reading, Laughing, and Connecting with Young Children

Rafael Ballagas, Joseph 'Jofish' Kaye and Hayes Raffle

Abstract In this chapter, we report on three projects that focus on storybook reading as a way to improve distance communication with very young children. "Connected Reading" builds on the insight that communication technologies for families with young children need to focus on *play* rather than *conversations*, and that having a shared activity can help structure this play. Our prototypes span a range of embodiments, from mobile video conferencing with physical books, to eBooks, and finally to video conferencing enhanced with depth camera technology. Our findings suggest guidelines to improve family communication with young children.

Introduction

According to the AARP[1], about half of grandparents live more than 200 miles away from their grandchildren (Davies and Williams 2002). How do families cope with this separation? In the summer of 2008, a group of researchers in Nokia Research

---

[1] The AARP is a non-governmental organization formerly known as the American Association of Retired Persons (see http://www.aarp.org).

R. Ballagas (✉) · J. 'Jofish' Kaye
Nokia Research Center,
IDEA Group,
200 South Mathilda Ave.,
Sunnyvale, CA 94086, USA
e-mail: tico.ballagas@nokia.com

J. 'Jofish' Kaye
e-mail: jofish.kaye@nokia.com

H. Raffle
Staff Interaction Designer,
Google, 1600 Amphitheater Parkway
Mountain View, CA 94043, USA
e-mail: hraffle@google.com

C. Neustaedter et al. (eds.), *Connecting Families,*
DOI 10.1007/978-1-4471-4192-1_9, © Springer-Verlag London 2013

Center Palo Alto began exploring how new tools for "Family Communication" could help families with young children maintain their relationships over a distance. We believed that young children and elders had the most time and desire to connect, but current technologies did not meet their needs. Our research goals were to understand the views and needs of long-distance families today, and to explore how new technology applications could help them form meaningful connections with each other.

Our work included field research and development of over a dozen technology prototypes. In this chapter, we report on three projects that use storybook reading as a way to interact with very young children over a distance. "Connected Reading" builds on the insight that communication technologies for families with very young children need to focus on *play* rather than *conversations*, and that having a shared activity like reading can help structure this play. Book reading is particularly successful because both the young and old understand and enjoy sharing books together, and the wealth of content makes it a rich playground for the young and old.

In the following sections, we will outline *Family Story Play*, *Story Visit*, and *People In Books*, three different embodiments of connected reading, and overview how each design makes long-distance interactions more playful, interactive, and fun for families to connect with young children over a distance.

Formative Research with Families

In order to understand the views and needs of American families today, we conducted qualitative studies with 22 diverse families in the San Francisco Bay Area between summer 2008 and spring 2009. These families were selected to span the spectrum of the Bay Area, including a variety of income levels, racial and ethnic identities, and occupations. Our original recruitment criteria were that the families included at least one child between the ages of 4 and 10; the realities of field studies meant that there were frequently siblings of a variety of ages present as well, giving us a pool that included many preschoolers as well.

In the first phase of the study we visited 18 families, of whom all used the telephone to communicate with their distant family members. Family visits followed a similar pattern: two to three researchers would visit a family's home at the end of the afternoon, when children would come home from school. We had the children take us on a tour of their room and show us their toys, which made them accustomed to our presence and meant we could observe them for the next few hours without them becoming shy. We would join the family for their evening meal, often bringing dinner with us, and we would also ask the family to schedule time to talk with a remote family member—nearly always a grandparent—with whom they often communicated. We would interview the parents in an open-ended manner about a variety of topics, including parenting practices, their attitudes to technology, toys and family, their values as a family, and their ways of learning about parenting. We video recorded interviews and took photos throughout the evening. Families were compensated for their time.

Interviews were later transcribed and coded using a variety of analysis techniques. Much of the content of many of the interviews was formally coding by two researchers using Atlas.TI. In addition, researchers read through transcripts, watched videos, listened to audio recordings, labeled, selected and reviewed pictures, and reinterpreted the results. Themes were discussed and thought through clustering sticky notes (in the manner of affinity analysis) and through shared brainstorming on whiteboards. Transcripts, video recordings and photographs were all placed on a shared drive accessible to the group, meaning that no one person held ownership or control over these materials. This enabled researchers to return to the source material at leisure to find illustrative photographs or quotes, as well providing opportunities for further reinterpretation and analysis at a later date (Kaye 2011). And, perhaps most importantly, these studies were interpreted through the act of creation of novel technological devices and experiences.

Family Communications *Phone Conversations with Children* (Ballagas et al. 2009) details the difficulties that families had in engaging with children over the phone. Many kids can't talk on the phone by themselves until 7 or 8 years old. Kids under this age have many cognitive, social, and motivational challenges that typically lead to communication breakdowns. For example, we observed one 3-year-old child during a call who repositioned the phone so that it was facing him and started kissing the speaker before clapping the phone shut, hanging up on the remote party. Clearly, he was really good at expressing himself physically through kisses and manipulating the physical affordances of the device by folding it shut. However, all of these expressions of love and action were lost on the remote party. While phones are accessible and ubiquitous, it is not obvious how to 'play' with someone over a phone.

After visiting the first 18 families and reviewing the transcripts, we noticed that two of the families were also using Skype or similar services to videochat with remote family in addition to telephone calls. We then recruited another five local families who used videochat. These families, along with the two from the original study, were the basis for our paper *Making Love in the Network Closet: The Benefits and Work of Family Videochat* (Ames et al. 2010). The procedure with these videochat families was more abbreviated than the other families, in that the visits were centered around a planned videocall with a remote family member (again, usually a grandparent) and subsequent interview. From the combination of work and the previously mentioned fieldwork we were able to build a picture of how the technically complicated and unreliable practice of videochat was a way for families to express their love and sense of identity as a family: *making love*, in the sense of creating and substantiating love—and creating and substantiating a sense of the family at the same time.

In our observations, video conferencing had clear benefits over telephone conversations in that it facilitated nonverbal communication: allowing children to show rather than tell, express through action instead of words, and use gestures and body language including 'skype kisses'. Families used video conferencing to include multiple parties, making it easier for parents to scaffold children in conversation. However, most families still had trouble keeping the children engaged for more than a few minutes because they primarily used videochat as an interface for conversation instead of play. In other words, videochat probably should be part of the solution, but videochat alone seems not to be sufficient for addressing families' desires for a sense of togetherness.

These visits and their associated study had number of ramifications to our research on Connected Reading. For example, nearly all families had difficulty keeping children engaged in communication, and it was clear this was an opportunity for design intervention.

This fieldwork led us to design a range of novel connected reading solutions to improve family communications. We hypothesized that providing a shared activity—in this case, reading a book together—would give structure to the communication and lead to longer richer interactions with young children. In our designs, we push current notions of books by adding novel interactive elements that bring the book to life and make reading more like play. We also hypothesized that there were opportunities for children to have meaningful learning experiences while engaging with long-distance loved ones.

Experiments in Connected Reading

Family Story Play

Family Story Play (Raffle et al. 2010; Ballagas et al. 2010) combines traditional paper children's books with an interactive agent (Sesame Street's Elmo) and mobile video conferencing. The system supports traditional reading experiences, including physical page turning, and is designed to fit into typical family rituals such as reading bedtime stories together. Family Story Play supports both "collocated reading" in which a copresent child can read the book with the child and play with Elmo, and "distance reading" in which a remote reader can be invited to read to the child over a videochat connection. When connected over a distance, the readers can see and hear each other through the video conference, and can also see what page the other reader is on. This is possible because each book is instrumented with small magnets to identify their current page, and sensors in the book frame can sense what page the reader is viewing. A remote reader's page information is displayed alongside their video image on the embedded screen (Fig. 9.1).

This project used a familiar children's character—in this case Sesame Street's Elmo—to engage children and adults in conversations with *each other* over a distance. Whereas muppets are typically given center-stage to entertain and educate children, we sought an opportunity where the muppet could engage the child and help both child and adult engage with each other. As such, we approached the mup-

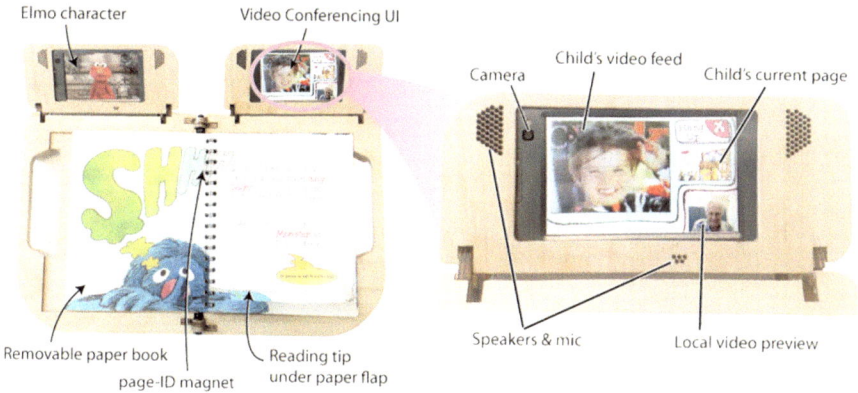

Fig. 9.1 'Family Story Play' allows families to read physical books together at a distance. The wooden housing holds two screens, one for video conferencing and a second for Elmo (who acts as a third member of the videochat). Sensors in the wooden housing detect the current page and update the remote reading partner's display

pets as teachers for both the child and the parent, whose role it was to engage the child and parent in educational dialogue together. Such dialogue around book topics is known to improve young children's literacy learning (Whitehurst et al. 1988; Zevenbergen and Whitehurst 2003), and in this system we showed that it can benefit family communication as well.

In Family Story Play, Elmo acts like a third party of the videochat. Video content of Elmo makes it appear as if he is listening to the adult as they read the story. Elmo models an interest in reading for the child. When prompted (by touching Elmo's screen), Elmo will ask questions related to the current page to inspire the child to talk more about the book. Research on literacy has shown that the more children talk about a book during a reading experience, the better their vocabulary development (Whitehurst et al. 1988), and Elmo can help young children learn in different ways. Adults may pose questions to Elmo and activate him to talk, making it seem like he is a part of the conversation. This can help with child engagement and enjoyment of the reading experience. Elmo can also provide scaffolding to remote readers: he asks children questions in the style of "dialogic reading" and can model for adults how to engage children in dialogue around book topics. To complement Elmo's role, we also provided simple text tips to support grandparents asking questions. Hidden under paper flaps on the book pages, adults could discover advice and suggestions for questions to ask young children about the book.

In a user study with eight families with children aged 2–4 (see Fig. 9.2), we compared reading with Family Story Play to reading a typical children's book over ordinary video conferencing. Our analysis showed that the shared activity of reading books seemed to be successful across both conditions. Even the traditional video conferencing with paper books had much more success in sustaining engaged communication with children compared to our fieldwork in which young children and

Fig. 9.2 Families reading together at a distance using Family Story Play. The system is designed such that the child reads with a collocated adult (*left*). The two devices connect using wireless LAN

adults tried to converse, and lacked an activity to organize their play. However, we coded videos for smiling and laughing and found that children and parents demonstrated significantly higher levels of enjoyment with Family Story Play compared to ordinary books. Why? Elmo was an important factor in keeping kids engaged, seemingly due to his star power with children. One parent commented, *"Elmo? She loved it. You saw her. She tried to kiss him."* (Father of 2.5 y.o. girl). This star power was not a clear positive; qualitative feedback revealed that some grandparents felt as if they might be in competition with Elmo for the child's attention. *"Oh I liked [Elmo]. I mean he brought up questions that I wouldn't even ask… He is a good influence, but when he beats me to the punch, that was a little distracting. [My grandson]'s not even looking at me or I mean—I don't know if he was even looking at the book. I think he might have been actually looking at Elmo over here, waiting for the ding or something instead of looking at the picture."* (Grandfather 3 y.o. Boy). These findings suggest that designers must strike a delicate balance when incorporating interactive characters into communication tools so that children's attention is directed in ways that are rewarding for all.

Also with Family Story Play, parents were twice as likely to give children control of the book pages (70 % of the time vs. 38 % of the time). We saw several instances of children engaging with pretend play during the reading experience, suggesting that Family Story Play helped children emotionally connect with their grandparents despite physical and technological barriers. These positive results encouraged us to extend the concept of connected reading to understand exactly what features of the system were most effective in helping families communicate.

StoryVisit

In order to evaluate our laboratory findings with a larger audience, we explored how connected reading might be brought to families anywhere in the world, for

Fig. 9.3 StoryVisit is a web-based embodiment of connected reading, allowing adults and children to read books together over video conferencing. The callouts illustrate parent tips (*top*), shared touch (*left*), and extended Elmo controls (*bottom*)

free. Our target audience was to provide the shared activity of book reading to families who were already engaging in family videochat with services like Skype. In 2010 we launched StoryVisit (Raffle et al. 2011), a prototype system that combined browser-based video conferencing and connected eBooks. The system included five titles from Sesame Street's ebook library, and built on learnings from our research with Family Story Play. By using digital instead of physical books, we are able to add new features to improve a sense of connectedness. In StoryVisit, pages are automatically synchronized; when the grandparent turns the page, the page also automatically advances for the child. (Either the adult or child may turn the page.) Furthermore, family members can point at objects in the book through *shared touch*—if one user points to the page, the remote user sees a large image of a hand appear in the same place (see Fig. 9.3). This allows children and adults to point to things in the book they are talking about facilitating nonverbal expression, as is particularly suitable for touchscreen tablet devices. Finally, with digital books, it is much easier to scale up the selection of books, and eliminates the issue of making sure both sides have the same physical book.

The design of Elmo was informed by Family Story Play, and kept many of the same elements. Elmo sits prominently in front of the book, drawing children to look at the book contents. Elmo can be controlled by the remote Reader using a menu of phrases that is not visible to the Child reader. This allows the remote Reader to invite Elmo into the conversations, prompting him to ask questions, or making him answer children's questions with a "laugh," a "yes" or a "no." Children may touch Elmo, causing him to do non-conversational things like laugh or dance (Fig. 9.3).

Like Family Story Play, conversation *Tips* are included for the remote Reader. They were displayed along the top of the book, and were not visible to children.

StoryVisit was launched publicly as a free service on the web (at http://www. storyvisit.org) in 2010. In the first 4 weeks, over 250 families registered to use the system, and 61 of them became 'active' users, using the system for at least one reading session with a long-distance reader, a 25 % uptake that reflects on the motivation and latent needs of this population to be better connected. In order to isolate the relative value of the Books, Elmo and reading Tips, families were randomly assigned to one of four different UI conditions: Elmo & Tips (similar to Family Story Play), Elmo Only, Tips only, and Book only (no Elmo and no Tips). Families completed an initial survey, and at the end of 6 weeks a post survey. Based on analysis of log data, a number of families were also invited to participate in telephone interviews about their experiences with the system. Finally, four of our families were treated differently from the start, in that they were explicitly recruited to use the system with heavy monitoring of usage. This included technical support and logging and analysis of video data. In total, our dataset included a wealth of quantitative and qualitative usage data about usage and satisfaction with the system.

Our results show that connected reading is significantly more successful than ordinary videochat for long-distance families to connect with young children. Families who used StoryVisit engaged in videochats with such young children for an average of 15 min with books alone, and an average of 21 min in the Elmo Only condition. This was a 5–8x increase over ordinary videochat durations observed in our formative research with Bay Area families who had young children, who usually sustained conversations with young children for only 2–3 min.

Significantly, usage of StoryVisit peaked for families with 3-year-old children, and total reading time for 3 year olds was significantly higher than for children under 3. Number of pages read was significantly higher than for children over 3. On the one hand, this peak of usage is expected since the book content was designed for 2–4 year olds. However, these findings are important because they mark the first ecologically valid data we know of that demonstrates that sustained distance communications with such young children is even possible.

Why did StoryVisit work with such young children? Data showed that content was key. The 'Elmo Only' condition performed significantly better than 'Book Only' in terms of average reading time per session and total reading time across all sessions. We were surprised that the 'Elmo Only' condition seemed to outperform the 'Elmo & Tips' condition. The data did not provide a clear cause—perhaps having both Elmo & Tips became overwhelming for users resulting in less interaction overall. Qualitative feedback conveyed the importance of Elmo in the design. "*I like the different choices and the fact that Elmo can ask comprehensive questions about things on each page. It would be great if he could have more than one question/ comment for each page. My son really liked to say, "Let's hear what Elmo says!" after his relative finished reading each page.*" (Family 75, 'Elmo Only' Condition).

Overall, the use of tips was very low. 75 % of the families in the 'Tips Only' condition clicked on a tip at least once, but tips were activated on only 7 % of all pages.

They sailed on a boat, a lovely boat. There was a sun in the sky. How far would they go

Fig. 9.4 'People in Books' depicts remote reading partners in the context of the story world alongside the characters

Use of tips was significantly lower when Elmo was present: in the 'Elmo & Tips' condition only 20 % of families clicked on a tip at least once, and tips were activated on less than 2 % of all pages. Although usage was low, some families in the 'Tips Only' condition found them valuable referring to them as *"[tips] have been 'how to be a good aunt' instructions… it's actually really helpful"* (Family 73, 'Tips Only').

In order to make connected reading sustainable for families with young children, it would likely need to be extended in several important ways. First, families expressed that they would like it to be part of their usual family videochat experience. As such, it should include ordinary videochat functionality like full-screen views. Furthermore, families wanted more content in the system. This would include larger libraries of eBooks as well as the ability to add personal content, such as existing favorite books. This type of personalization would likely expand usage of the system and allow the content to feel more personally meaningful.

People in Books

People in Books (Follmer et al. 2012; Follmer et al. 2010) immerses connected readers into the illustrations of a shared children's storybook. Through the use of custom depth camera, the system automatically removes people's background scenes from their video streams, allowing video of the child and remote reader to appear as if they are immersed in the storybook illustrations. Although the users are physically separated, People in Books uses videochat technologies to create the illusion that they are visiting a magical place where they can read and play together. Users' video images appear in surprising places, hanging from trees, hidden under covers, or sharing a boat ride with the story's main character (see Fig. 9.4). The goal is to encourage play and conversation about the book and to use the story "place" to create a sense of connectedness.

People in Books builds on some of the design principles learned from StoryVisit in that it helps a young child and remote adult connect over videochat with a connected eBook, and in that it uses interactive video to bring the book to life. Studies comparing reading experiences revealed that 'People In Books' is qualitatively different from systems like StoryVisit.

Children and parents felt closer together using People In Books. While using People In Books one mother commented, "*This one doesn't feel like we're separated. I feel like [I am] more close with Nicole.*" This sentiment was also exhibited in the way people used the system. One mother reached out towards her son in the book and said, "*I'm reaching out and grabbing you.*", to which the son responded, "*I can feel you*". This is a powerful example of how close people felt even though they were physically separated. We also saw instances of parents and children making kissing gestures and sounds towards each other on the screen echoing some of the physical expressions of love we saw in our earlier fieldwork. Other evidence of a strong sense of togetherness arose; for example one child needed a sense of security during reading, "*I can see a monster! Mama, Are you still next to me?,*" and both the mother and child leaned closer together in the story image. "*Now I am next to you, Mama.*" The mother responded, "*I'm going to protect you [from the monsters]*" and the child said "*Thank you mama!*"

We also saw more evidence of both sides engaging in pretend play using People in Books. For example, when one of the books depicted a river scene, one child lay on the couch and pretended to swim saying, "*I'm going to swim, swim, swim.*" Additionally, parents and kids would pretend to physically engage with the characters on the screen. One child acted as if he was snuggling up to the main character Max and said, "*I'm cuddling with Max.*" In another reading session, a parent pretended to tickle the feet of one of the monsters, making the far-away child laugh.

It seemed that immersing people's images into the same storybook illustrations achieved several effects. First, people were in a shared visual space, in contrast to the separate "windows" of typical videochat UI's. This created a sense of togetherness. Further, the playful illustrations and narratives encouraged children and adults to play together. There was a magic to "being there" with the story characters and the design seemed to support the kind of play that our early field work identified as a hallmark of successful distance communications with young children.

While the system seemed to offer many benefits for distance communication for families with young children, it still suffered from common pitfalls. Children would often hide or just disappear from the camera view because they do not always understand what the camera can "see." One parent commented that she had "*Less sense of what is going on in the room with People In Books.*" This may be a result of us not including a collocated adult with the children to ensure that children were in the field of view, and to articulate the child's actions for the remote adult to understand the context in the room. Despite these challenges, the project shows that advances in videochat technologies can support a greater sense of togetherness for families with young children through a combination of design and technology development.

Implications for the Understanding of Family Communication

Our fieldwork and exploration of novel connected reading experiences have brought us a deeper understanding into how to improve family communication at a distance and allowed us to generalize a few implications for design. In common with other authors in this book, much of our work is motivated by the need to connect young children with their remote grandparents (e.g., Moffat et al.'s chapter on Connecting Grandparents and Grandchildren in this collection). The following guidelines are further applicable to many different family relationships including traveling parents, divorced parents (e.g., Yarosh et al.'s chapter on this topic), or families dealing with long-term separation because of occupation (such as military families).

Create an Interface that is Fun and Facilitates Play One key lesson from our fieldwork is that you can't expect to have a conversation with a young child at a distance; instead you need to find a way to play with them. Although play through video conference can be challenging, our designs show a range of mechanisms that provide a playful shared activity. As designers, we should try to help families get technology out of the way so that they can play together.

Children Need Scaffolding As we saw in our trials, parents play a critical role in ensuring a smooth communication experience. Collocated parents actively articulated their children's actions and prompted them with questions to ensure that the remote partner understood the context on the child's side. When designing experiences for connecting families we need to consider how to better engage the collocated adult. Our designs currently lack an explicit role for collocated parents, which could impact adoption of these experiences over the longer term. Experiences will likely be most successful if they are designed to give collocated parents a clear role that is both enjoyable and rewarding.

Adults Need Scaffolding, Too Remote adults sometimes forget how to engage children, especially if they are not with the children on a day-to-day basis. Remote adults can also benefit from scaffolding and prompting to help them be more successful in engaging with children. Our designs used different kinds of scaffolding including the reading tips to encourage parents. In all of the designs, the reading activity scaffolded the interaction by giving remote adults and children something to talk about. In Family Story Play and StoryVisit, Elmo modeled dialogic reading techniques by asking open-ended questions about each page. We expect that with time, parents exposed to Elmo would be more likely to ask questions to children even when reading traditional paper books.

Allow for Personalization of Content Many parents expressed that content was one of the key reasons that motivated usage of a system. However, parents and children said that they wanted to be able to also read their favorite books. Expanding the library will help, and allowing families to scan and upload their own collections of books, images, drawings and personal mementos can be a different way of addressing this need.

Design for Offline Use Our fieldwork showed that many families had difficulty scheduling communication sessions with remote family members. In addition, many families expressed a desire to use these reading experiences at home, without a remote participant. We explicitly designed for offline use in Family Story Play, and designs should allow fluidly switching between collocated and remote reading activities in the same application.

Usability for Children Many of the children using StoryVisit's shared pointing feature tried to touch the screen directly. This was partly caused by our use of a hand image to convey the shared touch. However, this indicates that perhaps the design would be more successful if it was implemented on tablet hardware allowing for touching and swiping of the page instead of requiring interaction through the mouse.

There are Synergies Between Family Communication, Child Development, Emotional Expression, and Literacy Interaction with adults is key to helping children learn across a number of dimensions. Designers should remember that any interaction with a child is an opportunity for learning and growth.

Looking Ahead

With the emergence of social media on the Internet, our motivating questions are especially relevant today. Technology is creating new ways for people to connect, but most of today's tools still do not meet the needs of the young and old. Our research on Connected Reading shows that the combination of real-time communications channels with motivating content can help provide safe and compelling activities for families to engage in together over a distance. With Family Story Play we showed that books and children's characters can help children connect and learn from people they know and love. StoryVisit demonstrated that such systems can engage children as young as 3 years old, in the wild. And People in Books shows how people have a greater sense of connectedness by using Internet technologies to "travel to magical places" (like storybook worlds) together.

How will our research transition from laboratory studies and pilots to widespread tools that help families to connect more often and more successfully? One step is to begin developing products that address families as a group, and not just parents or children separately. Nokia, a company that does not market to children for ethical reasons, understood that families' needs—which include children's needs—could be met without treading into an ethically complex area of children's products. This can lead in a number of directions. For example, we are now working hard to commercialize some of our connected reading solutions. Our efforts are beginning with mobile eBook applications for collocated reading between a parent and child (see Fig. 9.5). 'Interactive Rich Reading' (Mori et al. 2011) maintains the interactivity

Fig. 9.5 Interactive Rich Reading is a mobile phone application that enables parents to read together with their children while collocated. Elmo is present to help bring the book to life

of the StoryVisit eBook design without video conferencing. The application allows for a parent and child to read together, and Elmo keeps young children engaged by bringing the book to life.

In the mobile devices marketplace, screen space is a limiting factor for designs like StoryVisit. As mobile devices become more powerful and capable, immersive designs like People In Books can be more successful. With larger touch screen devices becoming more prevalent, our work can change what "social media" means for families, for example by showing that families can share a story together to have a playful and educational experience over a distance.

The laugh of a child or smile of a loved one is what families treasure most—these are the experiences people want to have, to remember and to cherish. Connected Reading is a humble attempt to help families with young children to form connections over a distance. We hope to form a foundation for how companies like Nokia can get better at "connecting people" to the ones they love the most.

References

Ames, M. G., Go, J., Kaye, J. J., & Spasojevic, M. (2010). Making love in the network closet: the benefits and work of family videochat. *Proceedings of the 2010 ACM conference on Computer supported cooperative work* (pp. 145–154). New York: ACM.

Ballagas, R., Kaye, J. J., Ames, M., Go, J., & Raffle, H. (2009). Family communication: phone conversations with children. *Proceedings of the 8th international Conference on interaction Design and Children* (pp. 321–324). New York: ACM.

Ballagas, R., Raffle, H., Go, J., Revelle, G., Kaye, J. J., Ames, M., Horii, H., Mori, K., & Spasojevic, M. (2010). Story time for the 21st century. *IEEE Pervasive Computing, 9*(3), 28–36. IEEE Computer Society.

Davies, C., & Williams, D. (2002). *The grandparent study 2002 report*. Washington, DC: AARP.

Follmer, S., Raffle, H., Go, J., Ballagas, R., & Ishii, H. (2010). Video play: playful interactions in video conferencing for long-distance families with young children. *Proceedings of the 9th International Conference on Interaction Design and Children* (pp. 49–58). New York: ACM.

Follmer, S., Ballagas, R., Raffle, H., & Ishii, H. (2012). People in books: using a FlashCam to become part of an interactive book for connected reading. *Proceedings of the ACM 2012 conference on Computer supported cooperative work*. New York: ACM.

Kaye, J. (2011). Love, ritual and videochat. In R. Harper (Ed.), *The connected home: the future of domestic life*. London: Springer.

Mori, K., Ballagas, R., Revelle, G., Raffle, H., Horii, H., & Spasojevic, M. (2011). Interactive rich reading: enhanced book reading experience with a conversational agent. *Proceedings of the 19th ACM international conference on Multimedia* (pp. 825–826). New York: ACM.

Raffle, H., Ballagas, R., Revelle, G., Horii, H., Follmer, S., Go, J., Mori, K., & Spasojevic, M. (2010). Family story play: reading with young children (and elmo) over a distance. *Proceedings of the 28th international conference on Human factors in computing systems* (pp. 1583–1592). New York: ACM.

Raffle, H., Revelle, G., Mori, K., Ballagas, R., Buza, K., Horii, H., Kaye, J. J., Cook, K., Freed, N., Go, J., & Spasojevic, M. (2011). Hello, is grandma there? Let's read! StoryVisit: family video chat and connected e-books. *Proceedings of the 2011 annual conference on Human factors in computing systems* (pp. 1195–1204). New York: ACM.

Whitehurst, G. J., Falco, F. L., Lonigan, C. J., Fischel, J., DeBaryshe, B., Valdez-Menchaca, M., et al. (1988). Accelerating language development through picture book reading. *Developmental Psychology, 24*(4), 552–559. American Psychological Association.

Zevenbergen, A. A., & Whitehurst, G. J. (2003). Dialogic reading: a shared picture book reading intervention for preschoolers. In A. van Kleek (Ed.), *On reading books to children: parents and teachers* (pp. 177–200). Mahwah: Lawrence Erlbaum.

Chapter 10
Connecting Grandparents and Grandchildren

Karyn Moffatt, Jessica David and Ronald M. Baecker

Abstract Grandparent–grandchild relationships are diverse and ever evolving. Effective design of communications technology for them requires consideration of this complexity. This chapter considers grandparent–grandchild relationships from a life-course perspective, with the aim of identifying new opportunities for technology to support them. The grandparent–grandchild relationship is reviewed, discussing why it is important, identifying factors that challenge its success, and outlining its evolution over time. Current technology use is considered with the goal of identifying opportunities for improvement. A number of projects are presented as examples of the breadth of ways in which technology can support different grandparent–grandchild communication needs.

Introduction

> The child who reaches up to take her grandmother's hand as they cross the street will be different than the woman who reaches down 30 years later to again take her grandmother's hand as they cross the street, but they will still be holding hands. (Hodgson 1998, p. 183)

K. Moffatt (✉)
School of Information Studies,
McGill University,
3661 Peel Street, Rm 303C,
Montreal, QC, H3A 1X1, Canada

J. David · R. M. Baecker
Technologies for Aging Gracefully Lab,
University of Toronto,
Toronto, Canada
e-mail: jessicam.david@utoronto.ca

R. M. Baecker
e-mail: ron@taglab.ca

C. Neustaedter et al. (eds.), *Connecting Families,*
DOI 10.1007/978-1-4471-4192-1_10, © Springer-Verlag London 2013

Fig. 10.1 Percentage of individuals at age 20, 30, and 40 with one or more living grandparents, at select years over the twentieth century. Note the 10 year age shift over the century: roughly one fifth of 40-year-olds in 2000 and 30-year-olds in 1900, and three quarters of 30-year-olds in 2000 and 20-year-olds in 1900 had a living grandparent. (Graph based on data from Uhlenberg (1996))

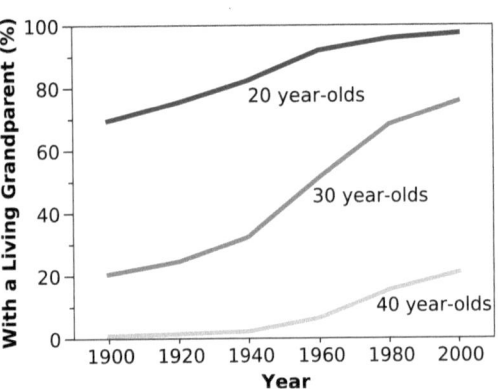

Rapid changes in life expectancy over the past century have dramatically altered the nature of grandparent–grandchild relations, with the result that they can now last well into the grandchild's adulthood. As shown in Fig. 10.1, three-quarters of 30-year-olds today have a living grandparent, as compared to one-fifth in the early 1900s (Uhlenberg 2004).

As designers of communications technology, these changes urge us to consider both the diversity across grandparent–grandchild relationships and the evolution within them. For example, the communication needs of a young child and his/her middle aged grandparent are very different a decade later when the child enters college and the grandparent reaches retirement, and different again when the child becomes a parent and the grandparent, a great-grandparent.

To date, most technology designed to support grandparent–grandchild interaction has focused on connecting young children with their grandparents (e.g., Davis et al. 2011; Khoo et al. 2009; Follmer et al. 2010; Judge et al. 2010b; Raffle et al. 2010; Vetere et al. 2009; Vutborg et al. 2010). As children are not proficient phone users before age seven, phone conversation—the currently dominant method of keeping in touch with long-distance family members—is difficult (Follmer et al. 2010), motivating researchers to seek other solutions, including those introduced elsewhere in this book (see chapters by Ballagas et al. and Judge et al.). When research has considered older grandchildren, it has mostly noted the challenges of communication with them but not offered many technological advances (Evjemo et al. 2004; Lindley 2011). Therefore, many aspects of the grandparent–grandchild relationship could also benefit from better communication media, especially as the grandchild ages.

In this chapter, our goal is to shed light on the nature of the grandparent–grandchild relationship from a life-course perspective and to identify ways to better support communication within it. In Sect. 2, we describe the grandparent–grandchild relationship in greater detail, discussing why it is important, identifying factors that challenge its success, and outlining its evolution over time. In Sect. 3, we provide an overview of ways in which current technology is used by grandparents and grandchildren with the goal of identifying points of failure and opportunities for improvement. In Sect. 4, we present five of our own research projects that support different grandparent–grandchild communication needs. Finally, in Sect. 5, we close with a discussion of recurring themes in this design space.

The Grandparent–Grandchild Relationship

There is clearly no singular grandparent–grandchild experience. Not all pairs will share a close relationship or desire greater contact, and within each pair, perceived and desired closeness will not necessarily be symmetric. We do not wish to suggest all relationships *are* or *should be* close. Rather, our overarching goal is to offer better opportunities for meaningful contact for those who want it. As such, we necessarily take an optimistic view of grandparent–grandchild relations and focus on identifying opportunities for technology to address unfulfilled needs.

Value and Significance

With this in mind, we can view the grandparent–grandchild relationship as one which offers an important source of mutual social support that is distinct from other family relations. For grandparents, it can be a source of joy and pride, and an opportunity to contribute to something meaningful, which helps create a sense of continuity and purpose (Peterson 1999; Kemp 2005). For grandchildren, the relationship provides an opportunity to establish deeper family bonds over an extended period of time. It is often a source of stability, mentorship, and encouragement.

Grandparent–grandchild relationships tend to be freer from conflict than parent–child relationships: grandparents are not typically responsible for (nor legitimately capable of) discipline, allowing them to more fully engage in nurturing and praise (Kemp 2005). For older grandchildren especially, grandparents can represent another "place to be"—a retreat from parents and siblings (Kornhaber and Woodward 1985). Moreover, close relationships with grandparents have been associated with lower depressive symptoms in late-adolescents and young adults, particularly in those from single-parent families (Ruiz and Silverstein 2007).

From their interviews with grandchildren, Kornhaber and Woodward (1985) identified five general themes in how grandchildren (aged 5–18) perceive grandparents. At the most fundamental level, grandparents were viewed as *nurturers*, providing love, shelter, protection, and nourishment. As a living ancestor, they were also seen as *historians*: curators of family stories, culture, and heritage, links to past generations, and windows into another time. In that sense, they were a source of knowledge to which the children had no other access, and presented an opportunity to imagine other ways of life. They were considered unusually positive *mentors*, a source of unconditional support and encouragement. They represented a diverse set of *role models*, providing exemplars for personhood, adulthood, grandparenthood, and seniorhood, among others. Finally, as playmates, younger grandchildren were often mesmerized by their grandparents' ability to manipulate the world, imbuing them with a *wizard*-like quality.

Adult grandchildren similarly view their grandparents as surrogate parents, buddies, storytellers, and confidants (Franks et al. 1993), though these roles evolve as the grandchild matures. Distinctly, the relationship becomes more equal in adulthood, with an increased focus on companionship (Kemp 2005). Though adult grandchildren do feel pressure and obligation towards their grandparents, these feelings are, for the most part, internally generated and are cast positively as "wanting to give back" in appreciation and respect for the older generation (Kemp 2005). Grandparents and adult grandchildren consider each other a resource that they can rely on as a "safety net" (Kemp 2005). Adult grandchildren often cherish the values, lessons, and beliefs they acquire from their grandparents, and there is some evidence that grandparents also learn and adjust their values based on interactions with their grandchildren (Seponski and Lewis 2009).

Finally, there are also broader benefits. Close grandparent–grandchild relations provide a special opportunity for developing cross-generational understanding. For most people, it is their longest lasting relationship with someone from a nonadjacent generation, and as such, the grandparent–grandchild relationship forms the principle place where intergenerational competencies are learned (Harwood 2000b). The work of Kornhaber and Woodward (1985) illustrates this point nicely. Their interviews with 300 grandchildren (aged 5–18) revealed that those with at least one close grandparent were less likely to fear old age and more likely to view older adults positively, than those who did not. Thus, the grandparent–grandchild relationship has important significance both for grandparents and grandchildren themselves, and for society as a whole.

Reciprocity of Support

Human computer interaction research concerning older adults[1] often focuses on the ways in which technology can compensate for cognitive and sensory impairments to enable individuals to live more independently (e.g., Hawkey et al. 2005; Lee and Dey 2007; Mynatt et al. 2000; Rowe et al. 2007; Wu et al. 2007). This body of research addresses real and important needs, and there is no doubt that many older adults do have significant impairments and need substantial support. However, when this work is read without a broader understanding of how these impairments fit within the general context of aging, it can unintentionally bias readers towards primarily viewing older adults as support recipients.

---

[1] We acknowledge that not all older adults are grandparents, and not all grandparents are older. However, the two groups overlap sufficiently for the purposes of this discussion: as of 2001, nearly 75 % of Canadians 65 or older and less than 2 % of those 45 or younger were grandparents (Turcotte and Schellenber 2006).

In reality, many grandparents today give more support than they receive (Hoff 2007), and this has a positive affect for most older adults (Keyes 2002). They are healthier, better educated, and more financially secure than any group of elders before them (Uhlenberg 2004). As such, they are have more time and energy to devote to their families, and correspondingly require less financial or caregiving support from younger generations, or require it much later. Moreover, declining fertility rates have led to fewer cousins and siblings competing for grandparent attention and less overlap between parenting and grandparenting roles. Thus, today's grandparent typically has more capacity to provide (social, emotional, and financial) support, and with fewer grandchildren vying for it, each child stands to receive more.

Evolution Over Time

Beyond the general themes outlined thus far, the grandchild's view of grandparents evolves substantially over time, as their needs and perceptions, and correspondingly their expectations, change (Kahana and Kahana 1970).

Up to about age five, children view their grandparents as additional parents, valuing them for the love, attention, and presents they provide. As the child gets older (ages 8–9), the balance shifts. The pair become more like companions or playmates, and the relationship becomes more reciprocal, with a focus more on "doing together" than "providing for." However, this golden period is often followed by an abrupt shift as the child enters the pre-teenaged years during which children typically distance themselves from family as they seek independence.

The relationship begins to regain some solidarity as the grandchild enters the late teens. As teenagers get older, they again place more value on their relationships with grandparents. Both Hartshorne and Manaster (1982) and Robertson (1976) found that the majority of their teenaged participants held positive attitudes about spending time with grandparents. Dellmann-Jenkins et al. (1987) found that teenagers viewed grandparents as confidents with whom they could discuss personal issues. Thus, even if teenaged grandchildren do not appear to seek closeness with their grandparents, it is important not to underestimate the value they place on them, and the comfort they find in having them as an available resource. Teenagers can also begin to develop a sense of responsibility towards their grandparents. In observing teenagers with institutionalized grandparents, Streltzer (1979) found they were highly concerned with wanting to know what they could do for their grandparent.

The transition to adulthood also brings about evolution and change in the relationship. Notably, it marks a move to independence and away from parental mediation. It is also a time when major life transitions to college and career can result in increased geographic separation and competing responsibilities. These changes can make it difficult to sustain contact (Sheehan and Petrovic 2008).

Finally, we note that these stages build upon one another. Developing a close grandparent–grandchild relationship in childhood is especially important as it sets the stage for a solid relationship in adulthood (Geurts et al. 2011).

Additional Factors Impacting the Relationship

Given the complexity of the grandparent–grandchild relationship, it is not surprising that it can be influenced by a wide variety of internal and external factors. We cannot fully cover them here, but we briefly highlight those which are particularly relevant to designers of communications technology.

Life achievements such as employment, parenthood, and marriage can all affect intergenerational solidarity, but the direction and magnitude of their impact is not always straightforward (Mills 1999). These roles can be sources of commonality that bring grandparents and grandchildren together, but they can also be points of divergence that strain the relationship. For example, employment attainment can bring together a grandson and his grandfather by giving them something in common, but can alienate a homemaker grandmother from her career-oriented granddaughter. Similarly, birth of a child can promote solidarity by fortifying interest in family ties; however, it can create a divide if there are differing view points on parenting and child rearing (Glass et al. 1986).

Divorce can be a particularly powerful force on grandparent–grandchild relations. Parents of the custodial parent may see their role grow, providing them with increased opportunities for contact and closeness, particularly if they are called upon to help out with parenting. However, for parents of the non-custodial parent, it can cause excessive strain (Kornhaber and Woodward 1985). Sometimes grandparents in this situation lose contact because the custodial spouse moves away, but even if they are geographically proximate, the social strain between grandparent and ex-child-in-law can drastically impede contact between grandparent and grandchild. Supporting these individuals may be a particularly fruitful opportunity for designers and researchers to explore.

Retirement as an extended period of one's life without work and with limited responsibility is a relatively new concept, dramatically impacting grandparenthood. In particular, moving away—typically to a warmer climate, afar from work or "busy life"—has become not only acceptable over the past few decades, but representative of an ideal. Interviews with grandparents who had moved to Florida for retirement (Kornhaber and Woodward 1985), revealed that many had not anticipated the impact moving would have on their family relations, especially those with grandchildren. Instead, they were surprised—and disappointed—to discover that they were unable to live up to the image of grandparenthood they held from their own childhood. Though these grandparents were cynical about repairing their relationships, feeling that it was "too late for them," it seems likely that these grandparents would have been interested in opportunities for sustaining contact had they existed.

Gender, kinship, and ethnicity also influence grandparent–grandchild relations. Women tend to be closer to their grandparents than men, grandchildren tend to be closer to maternal grandparents (particularly maternal grandmothers), and different cultures and ethnic backgrounds place different emphasis and values on the grandparent role (Sheehan and Petrovic 2008).

In sum, all of these factors are potentially important to the design of communications technology. It is particularly important to consider them when conducting research on grandparent–grandchild relations or evaluating potential designs to ensure that sample-specific findings are not inappropriately generalized.

Communication Media Use Today

Family studies literature has expanded in recent years to explicitly address technological support for grandparent–grandchild communication. One common characteristic of this work is that it tends to capture only a particular stage in the relationship: connecting grandparents with young grandchildren. While it is unlikely that a single technology can meet the ever-evolving needs of grandparents and grandchildren, it is important to explore the ways technologies fit into particular stages and cover different kinds of relationships. Though many factors can challenge or impede this relationship, we focus on those which inhibit face-to-face interaction, as they seem most ripe for technological intervention.

Telephone

The telephone is the standard tool for long-distance communication, yet it is also one fraught with many pitfalls. Children up to the age of nine can have difficulty engaging in phone conversations (Ballagas et al. 2009), and though older children and teenagers have sufficient phone skill, their calls are as infrequent, as short, and as likely to be parent-initiated as those of younger children (Evjemo et al. 2004). Though there is a general tendency for teenagers to distance themselves from family relationships, telephone phone calls seem particularly problematic. Evjemo et al. attributed this to the phone providing insufficient support for developing a conversational context. When interacting face-to-face, grandparents and grandchildren participate in a wide variety of activities with one another, including watching TV, playing games, and going on outings (Dellmann-Jenkins et al. 1987). These activities provide a shared experience that can be used as the basis for conversation, and this grounding is missing from phone communication.

The telephone can likewise be difficult for older adults. Phone calls often arrive unexpectedly, which can be challenging for older adults with cognitive deficits as they are less able to plan for the conversation to compensate (Ryan et al. 1998). Moreover, individuals with hearing loss cannot use visual cues to compensate for auditory decline, and these challenges can be compounded when the grandchild over-compensates for the grandparent's deficits, which can be seen as patronizing (Harwood 2000a). Chronic pain can also be a barrier to phone use as sustained periods of holding the phone can be uncomfortable and challenging for these individuals (Benjamin et al. 2012). In general, the form factor of some technologies can make it painful or cumbersome for those with a physical disability to use. Designers of technology should consider how to balance the richness of synchronous communication with the issues surrounding phone calls.

Email

The second most popular form of communication media is email (Dickinson and Hill 2007; Tee et al. 2009). Though email is often perceived by older adults as lacking the personal touch of a phone call or handwritten letter (Lindley et al. 2009), a number of strengths mitigate this limitation.

Email covers long distances and time zones in ways that phone calls and letters cannot (Lindley et al. 2009). For example, a grandparent living in Toronto can send an email late in the evening to a grandson living in London, who can respond early the next morning. This examples highlights two advantages: (1) email can be useful when a quick response is needed or desired, and (2) it enables the sender to initiate communication without interrupting or disturbing the recipient.

An additional advantage of email is that it enables easy sharing of digital content. A number of studies have documented the sharing of digital photos over email (Frohlich et al. 2002; Kirk et al. 2006; Miller and Edwards 2007; Tee et al. 2009), noting that these exchanges can serve as the basis for subsequent conversation. It is possible that a fluid conversation over email might be preferable to a stilted one over the phone, but such nuances have not yet been explored.

Finally, the informal nature of email, though typically disliked by grandparents, can be appealing for grandchildren (Dickinson and Hill 2007). Indeed Harwood (2000a) suggests that low-richness media like email may be ideal from the grandchild's perspective, specifically because it is less personal. Thus, there is a clear tradeoff: older adults must balance a desire for more intimate contact, with the likelihood of it being less frequent.

Video Chat

Video calls (using programs such as Skype and iChat) are becoming increasingly popular for face-to-face conversations over a distance. Kirk et al. (2010) and Judge et al. (2010a) both examined the adoption of video-mediated communication in the home setting, broadly capturing adoption patterns across different relationships (e.g., teenager–friend, adult child–parent, and distance-separated couples). Interestingly, both studies observed the use of "open connections," a practice of leaving a video connection open for several hours to enable a sense being together without continuous conversation or attention.

Ames et al. (2010) specifically studied the use of video chat to connect remote grandparents to young grandchildren (and their parents). They found that young children had varied levels of participation during video calls and were more engaged than they typically are during phone calls. For grandparents, video calls provided an increased opportunity to see grandchildren, giving them a sense of "being there." However, there were challenges: web cams and chat programs need to be properly installed, an appropriate time for the call needs to be arranged, and each party needs to "prepare the scene" before the call.

Despite these drawbacks, the many advantages of video chat have motivated designers to leverage low-cost video-conferencing applications to support play, learning, and collaboration at a distance between grandparents and young children. Ballagas et al., in their chapter on reading, laughing, and connecting with young children, explore a number of systems that enable grandparents and grandchildren to read together over a distance (see also Raffle et al. 2010, 2011; Follmer et al. 2012). Similarly, work done by Vutborg et al. (2010) enables grandparents to tell fictional stories over video chat, while photos taken by the grandchild inspire discussion about current happenings in the child's home. Always-on technology, such as the Family Portals work described in Judge et al.'s chapter on private and public messaging (see also Judge et al. 2010b), provide a continuous peripheral connection between homes that, similar to the "open connections" described above, provide a window into a grandchild's life that may not be captured by scheduled calls.

Moving Forward

In sum, the past few years have brought a great deal of progress in terms of connecting grandparents to their grandchildren using technology. Recent efforts have built a solid understanding of many challenges inherent to intergenerational communication; however, we do not yet have a solid grasp of how to bridge conflicting needs and preferences. To address the needs and preferences of younger family members, a number of researchers have proposed lightweight mechanisms for staying in touch (e.g., Lindley 2011; Mynatt et al. 2000; Romero et al. 2007; Tee et al. 2009). Unfortunately, older adults typically desire richer contact than these interactions provide (Lindley 2011), and it is unclear how this conflict should be reconciled.

Thus, many opportunities for innovation and development remain. We especially see promise in approaches than merge asynchronous and synchronous components to enable fluid negotiation (and renegotiation) of desires and capabilities. Also promising are methods that support asymmetric participation, thereby allowing individual flexibility in the quantity and composition of participation.

Moreover, additional needs remain to be investigated. In particular, research has tended to focus on grandparents who live at home (and indeed on enabling them *to* live at home). Much less attention has been placed on communications media use with families where the grandparent is in an institutional setting or nursing home. Gaver et al. (2011) have begun to explore the distinct and interesting problems that arise is this unique environment. We encourage researchers and designers to continue work in this vein.

Supporting Diverse Grandparent–Grandchild Bonds

In our work, we have begun to explore different aspects of grandparent–grandchild relations, with the goal of fostering deeper connections across all its stages. In this section, we present five projects that illustrate the broad range of relations that can be supported. These research projects in no way address all the needs of grandparents and grandchildren but rather represent a starting point, which we hope will inspire future endeavors.

The first two projects, Take Me With You and Shared Stories, are in the early stages of development, but most explicitly target grandparents and grandchildren. Both aim to create an activity that grandparents and grandchildren can share remotely but in two unique ways: Take Me With You focuses on collaborative play and connecting with young children, while Shared Stories targets young adult grandchildren and uses family history to support interaction. The remaining three projects, were all designed from the perspective of supporting specific older adult needs: the ALLT-book project supports older adults with print-disability by enabling collaborative reading, Families in Touch supports those with chronic pain via a communicating picture frame, and Multimedia Biographies helps individuals with dementia to engage in conversations about their past. However, in doing so, each also provides an opportunity for grandparent–grandchild interaction; our goal in this section is to draw out those opportunities.

Take Me with You: Remote Intergenerational Play

Our concept for Take Me With You is a shared adventure game that promotes physical activity, cognitive stimulation, and social engagement, by using these elements to move the narrative of the game forward. Seniors partner with their grandchildren to play together even when they are not in the same place or time. Take Me With You is currently under development as a proof of concept game for the iPhone and iPod Touch.

To illustrate the game, consider the fictitious example of 8-year-old Lucy and her 70-year-old grandmother Vivian. Lucy and her grandmother are close but their visits are infrequent since Lucy and her parents moved from Montreal, Quebec (where Vivian still lives) to Portland, Oregon. Lucy and Vivian have thus started playing Take Me With You to stay in touch between face-to-face visits.

Movement through the Take Me With You world (Fig. 10.2a) is fuelled by physical activity: both Lucy and Vivian move in the real world to progress through the virtual map. Because this physical activity is designed to be flexible, Lucy advances her character by running around in her backyard and local park, pretending it is the imaginary world, while Vivian, who has trouble getting out during the cold winter months, advances hers by walking up and down her apartment hallway. A soundscape supports eyes-free interaction and sets the scene by receding and advancing as Lucy and Vivian leave and enter landmarks on the map.

Fig. 10.2 Take Me With You uses **a** pedometer-driven gameplay to encourage physical fitness, **b** mini-games to provide appropriate cognitive stimulation, and **c** digital treasures, such as photo collages, to promote social engagement

As they explore the virtual space, the pair encounters challenges such as word games and brainteasers that are designed to promote appropriate cognitive stimulation for both age groups (Fig. 10.2b). Successful completion of these challenges earns digital treasures, such as photo collages or collaborative spoken stories. These are intended to create lasting artifacts that represent the time spent together and encourage feelings of closeness. Figure 10.2c provides an example of a photo collage treasure. Vivian earns the reward first and is asked to take a picture of herself. Later when Lucy completes the challenge her photo is added and they both receive a copy of the completed collage, which can be printed or shared with others.

Reflecting back on the grandparent roles identified earlier in the chapter, Take Me With You primarily draws on the playmate role, and correspondingly, it chiefly targets grandchildren aged 7–10. It also touches on the theme of building and developing family roots; the virtual treasures can become a shared keepsake, lasting well beyond interest in the game.

Shared Stories: Connecting with Family History

Our Shared Stories concept aims to address the gap between the lightweight communication mechanisms favored by young adults and the rich contact desired by

Fig. 10.3 The Shared Stories concept. **a** The grandchild manages the photo-story book, scanning in photos and organizing the content. **b** To collect stories, the grandchild attaches short audio messages to photos, which are sent to the grandparent via a digital picture frame. **c** The grandparent uses a digital pen to link audio and handwritten stories to pictures, and sends them back

older adults, using the construction of family history archives as a shared collaborative activity. Prototypically, we envision asymmetrical use with the grandchild taking charge of constructing and organizing the digital picture book, and the grandparent providing the stories and content. As the grandchild scans and organizes the photos, s/he can select photos that are unfamiliar or representative of an interesting event and attach an audio message such as "Who's in this photo?" or "Tell me more about this day?" (Fig. 10.3a). The audio recording and photograph are sent to the grandparent via a wireless picture frame (Fig. 10.3b), and the grandparent responds with an audio or handwritten story (Fig. 10.3c) using a wireless digital pen such as the Livescribe Connect[2] and a specially designed diary to record and send the story back to grandchild.

Our choice to limit communication to asynchronous pre-recorded messages and handwritten stories is intentional. We chose audio-recordings because they are more personal than a short text snippet, which we predict will be appreciated by grandparents. Coordinating synchronous discussion may be troublesome to young adult grandchildren who want to work in short bursts or at odd hours, or who may fear "getting trapped" in longer than planned conversations, motivating an asynchronous design. Though older adults often dislike lightweight exchanges such as those encouraged by asynchronous communication, we hypothesize that they may find these particular ones more meaningful as they reflect effort invested in a shared project. We chose to use handwritten and audio stories both because they reduce the technical demands on the grandparent, and because they personalize the digital archive. Ultimately, digital replicas of handwritten stories and recordings of the audio stories are embedded in the photo book.

Shared Stories is currently in the early stages of design. Many of the design choices presented here reflect early findings from a survey of older adults' perceptions of communication technology and family history archiving practices; analysis of this data is currently underway.

---

[2] http://www.livescribe.com.

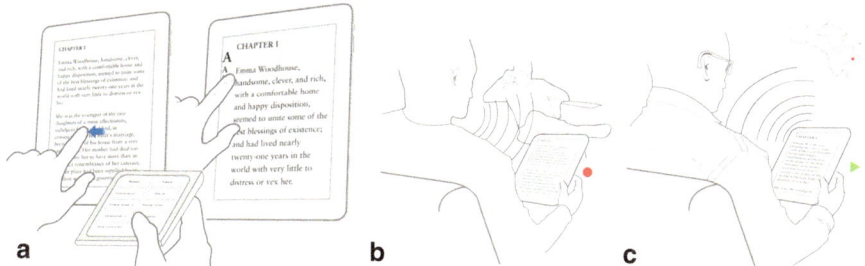

Fig. 10.4 The ALLT-book. **a** It supports alternative input devices such as a keypad, making it *accessible* to people with different sensory and motor abilities, and provides easily accessed *large-print* magnification for those with low-vision or who find reading larger text easier. **b** It *listens* by recording the voice of a friend, family member, or caregiver reading to the user, and **c** *talks* by speaking the text aloud using ether text-to-speech or a previous personal recording

ALLT-Book: A Collaborative Reader

Having a book read aloud is a common way of reading for many individuals with a print disability.[3] It is likely especially important for those who acquire one later in life, as they are less likely to master an alternative such as braille (Douglas et al. 2006). Collaborative reading, however, has historically been an ephemeral experience, only available while a reader is present.

The ALLT-book, shown in Fig. 10.4, is an iPad application that makes this content persistent by recording the audio of a reading, storing it synchronized with the text,[4] and making it available to the print-disabled user through an accessible interface (Snelgrove and Baecker 2010). Within a family context, the ALLT-book provides more than just access to print materials: it provides an opportunity for meaningful interaction. Over time, the recordings may become a cherished reminder of the time spent together. The interface could easily be extended for remote use; for example, imagine an adult grandson reading the morning news for his grandmother before work each day, preparing it for when she gets up in a later time zone.

Though not specifically targeted to grandparents and grandchildren, the ALLT-book is an example of the kinds of technology we believe can facilitate grandparent–grandchild interaction. As a shared-activity, it provides the kind of support identified by Evjemo et al. (2004) and Vutborg et al. (2010) as crucial for successful grandparent–grandchild interactions. It also provides an opportunity to provide

---

[3] Print disability includes a broad spectrum of visual, perceptual, and physical disabilities, including sight impairments, learning disabilities, and any other cognitive or physical disability that prevents a person from reading a standard print edition of a book. In Canada, its prevalence is estimated to be 1/10, increasing with age (Canadian Library Association 2005).

[4] Currently, this synchronization is achieved at the sentence level by having the reader gesture as they advance through the text. We are also exploring the use of natural language processing techniques to automate this task.

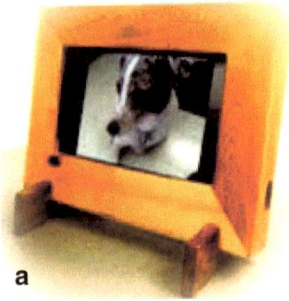

Fig. 10.5 The Families in Touch Frame. When the frame is touched, an email is automatically sent to loved ones asking them to reply with a video; once a new video message has been recorded it is sent back to the frame for remote viewing. **a** The first version, consisting of a netbook encased in a wooden picture frame. **b** The second version on a tablet computer

support, an objective of many older grandchildren (Streltzer 1979). In a recent pilot evaluation, a blind woman in her late 70s used an ALLT-book with her college-aged grandson over several sessions. She was able to master the touchscreen interactions (though a few refinements to the interface were identified), commented positively on the quality of the recordings, and generally enjoyed the experience. A field evaluation of the refined ALLT-book prototype is currently underway in the home of a 40-year-old educated woman with MS who has not been able to read a book for 10 years. Early results are encouraging: she is happily reading together with her family.

Families in Touch: A Communicating Picture Frame

To enhance family connections with older adults, a number of research projects have sought to leverage picture frames as a natural focal point in the home. These projects have mostly focused on augmenting photos with sensor data to support better awareness of activity (e.g., Mynatt et al. 2001; Consolvo et al. 2004). We build on this work but take a slightly different approach by instead using picture frames as a portal for accessible video chat communication. Our Families in Touch communicating picture frame consists of a touch screen computer fitted inside of a wooden frame (Fig. 10.5). When the frame is touched, an email is sent to loved ones, encouraging them to log on to a web site to upload or record a video for the frame owner. The videos are then sent back to the frame, and once the new content has arrived, the user touches the frame again to view it.

Using data from interviews and a pilot deployment study (Benjamin et al. 2012; David et al. 2011), we designed Families in Touch to address the unique communication needs of older adults with chronic pain, which is defined as pain that persists after an injury has healed or as pain that lasts longer than 6 months (Gatchel et al.

2007). Chronic pain carries with it significant social stigma, misunderstanding, medical disbelief, and barriers to finding appropriate treatment (Clarke and Iphofen 2008); thus, isolation can be a prominent feature of having chronic pain. Social isolation has its own medical consequences (Litwin 1998), compounding the effects of chronic pain. Pain is often intermittent and variable, which can make it difficult to plan interactions and to sustain long conversations such as face-to-face visits, phone conversations, or video chat sessions (Benjamin et al. 2012; David et al. 2011). Evidence suggests that increased social contact and support can promote positive health status (e.g., Tomaka et al. 2006; Jamison and Virts 1990). Thus, we designed Families in Touch as a minimal effort avenue for those with chronic pain to reach out and receive rich contact from loved ones.

Though the system was designed to meet the needs of those with chronic pain, its asymmetric and asynchronous design may also be a good fit for older grandchildren. The email requests can help remind teen- and college-aged grandchildren to provide contact, while their short and impersonal nature should limit the pressure on both the grandchild (of a perceived obligation to respond) and the grandparent (of a need to respect boundaries). The video responses also seem a good fit for older grandchildren as they allow for control over timing and duration. The second version of the system (Fig. 10.5b) is currently being designed and will be followed by a deployment study.

Multimedia Biographies: A Catalyst for Conversation

In this project, we worked with individuals with mild cognitive impairment and family members of individuals with Alzheimer's disease to create multimedia biographies from family photos, home movies, documents, music, and narration (Smith et al. 2009; Damianakis et al. 2009). When viewed, the multimedia biographies helped participants reminisce about their past and engage in conversation around life stories as shown in Fig. 10.6. Family members and participants perceived the biographies as a means for preserving personhood, helping third-party caregivers to better connect with the participants, and preserving the participant's story for future generations.

With respect to grandparent–grandchild interaction, multimedia biographies can help support face-to-face interaction by serving as a conversational support, easing the pressure of finding a conversation topic. Because the biographies are inherently personal, they can prompt further sharing, and thereby, act as a catalyst to conversation. Multimedia biographies could also be constructed earlier, prior to cognitive decline, providing an opportunity for grandparent–grandchild collaboration. Producing multimedia biographies can be a time-consuming process, requiring technical savvy. Young adult grandchildren may currently be best suited for taking on the technical aspects of making the biographies, while the grandparents themselves are best suited to shaping the content.

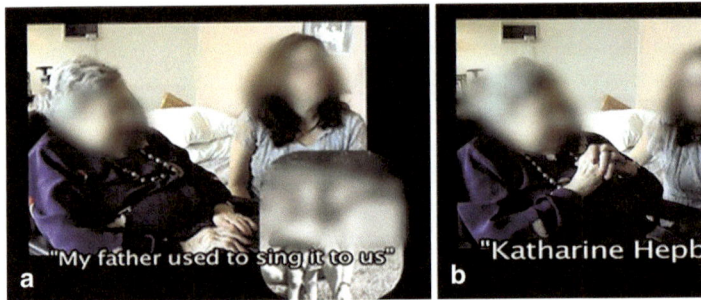

Fig. 10.6 Multimedia biographies. Participants and family members who watched the videos together often engaged in conversations around life stories

Conclusion

This chapter has described grandparent–grandchild relationships with the goal of illustrating their breadth, diversity, and evolution over time. A few themes stand out, which we discuss here.

The notion of asymmetry recurs in both the literature and many of our own projects. Lindley (2011) provides an overview of how asymmetry has been observed in a number of family communication projects, including her own. We propose leveraging this natural asymmetry. For example, with Families in Touch the older adult can only send a precomposed message to a fixed group, but respondents reply with a personalized and rich video. Shared Stories imposes a different type of asymmetry by assigning different roles to the grandparent and grandchild.

The tension between synchrony and asynchrony is also compelling. Though there is some evidence that older adults may prefer the prolonged contact offered by synchronous communication, asynchronous communication offers certain advantages for intergenerational exchanges: it can accommodate busy competing schedules, provide control to each party over how much time and effort is dedicated to the exchange, and enable respondents to reflect on their communication before sending it (Lindley 2011). Both Families in Touch and Shared Stories impose asynchronous interaction, while ALLT-book and Take Me With You support both. Multimedia Biographies primarily encourages synchronous interaction, though their production could introduce asynchronous elements.

Supporting collocated interaction between grandparents and grandchildren has been relatively under-explored. Most often, family communication technology is considered as a means of bridging geographical separation. However, two of our projects, ALLT-book and Multimedia Biographies, primarily support collocated interaction. Because intergenerational interaction is so deeply entrenched in shared activity, we believe that there are many opportunities for supporting collocated interaction, particularly since technology can provide an opportunity for children to offer expertise, partially balancing power in the relationship (Aarsand 2007; Voida and Greenberg 2009).

In closing, we return to the grandparent roles identified by Kornhaber and Woodward (1985): nurturer, historian, mentor, role model, and wizard. Some of these roles also appear in technology design. In particular, the role of wizard or playmate appears frequently in communications technology designed for grandparents and grandchildren (e.g., Davis et al. 2011; Judge et al. 2010b; Follmer et al. 2010; Lindley 2011; Raffle et al. 2010, 2011; Vetere et al. 2009), and a few projects have additionally leveraged the role of family historian or storyteller (e.g., Raffle et al. 2010, 2011; Vetere et al. 2009; Vutborg et al. 2010). The emphasis placed on play reflects back to our early observation that most design effort has focused on the needs of younger grandchildren. In designing for older grandchildren, the remaining roles of nurturer, mentor, and role model offer new design avenues to explore.

Acknowledgments   The research projects described in this chapter reflect the work of students and research assistants supervised by Dr. Ronald Baecker in the Technologies for Aging Gracefully Lab (TAGlab). *Multimedia biographies* was a joint research project carried out by a team led by Masashi Crete-Nishihata and Karen L. Smith. *Families in Touch* was conceived by Elaine Macaranas for her undergraduate thesis at the Ontario Collage of Art and Design, undergraduate student Thariq Shihipar helped build the initial prototype, and the project is currently led by Jessica David for her master's project. *ALLT* was initially designed and build by Xavier Snelgrove for his undergraduate thesis, with input from Bev Dywan, Greg Van Alstyne, Leila Rezai, Velian Pandeliev, and Karyn Moffatt. *Take Me With You* is led by Deborah Ptak, with development support from Nermin Moufti, Nick Shim, and Sarah Strong. *Shared Stories* is Karyn Moffatt's postdoctoral project. We would also like to thank the rest of TAGlab for their feedback, and to especially thank Garry Ing for creating artwork. TAGlab is grateful for financial support from NSERC, GRAND NCE, CIHR-HCTP, Microsoft Research, Google Research, MyVoice, the Alzheimer's Association, and the Connaught Fund of the University of Toronto.

References

Aarsand, P. A. (2007). Computer and video games in family life. *Childhood, 14*(2), 235–256.

Ames, M. G., Go, J., Kaye, J. J., & Spasojevic, M. (2010). Making love in the network closet: the benefits and work of family videochat. *CSCW'10: Proceedings of the ACM Conference on Computer Supported Cooperative Work* (pp. 145–154). New York: ACM.

Ballagas, R., Kaye, J. J., Ames, M., Go, J., & Raffle, H. (2009). Family communication: phone conversations with children. *IDC'09: Proceedings of the 8th International Conference on Interaction Design and Children* (pp. 321–324). New York: ACM.

Benjamin, A., Birnholtz, J., Baecker, R., Gromala, D., & Furlan, A. (2012). Impression management work: How seniors with chronic pain address disruptions in their interactions. *CSCW'12: Proceedings of the ACM Conference on Computer Supported Cooperative Work* (799–808)

Canadian Library Association. (2005). *Opening the book: a strategy for a national network for equitable library service for Canadians with print disabilities [Working Group Report].* Ottawa: Canadian Library Association.

Clarke, K. A., & Iphofen, R. (2008). A phenomenological hermeneutic study into unseen chronic pain. *British Journal of Nursing, 17*(10), 658–663.

Consolvo, S., Roessler, P., & Shelton, B. E. (2004). The CareNet display: lessons learned from an in home evaluation of an ambient display. *UBICOMP'04: Proceedings of the 6th International Conference on Ubiquitous Computing* (pp. 1–17). Nottingham, England.

Damianakis, T., Crete-Nishihata, M., Smith, K. L., Baecker, R. M., & Marziali, E. (2009). The psychosocial impacts of multimedia biographies on persons with cognitive impairments. *The Gerontologist*, 50, 23–50.

David, J. M., Benjamin, A., Baecker, R. M., Gromala, D., & Birnholtz, J. (2011). Living with pain, staying in touch: exploring the communication needs of older adults with chronic pain. *CHI EA'11: Extended Abstracts of the SIGCHI Conference on Human Factors in Computing Systems* (pp. 1219–1224). New York: ACM.

Davis, H., Vetere, F., Gibbs, M., & Francis, P. (2011). Come play with me: designing technologies for intergenerational play. Universal Access in the Information Society, Online First June 2011. doi:10.1007/s10209-011-0230-3.

Dellmann-Jenkins, M., Papalia, D., & Lopez, M. (1987). Teenagers' reported interaction with grandparents: exploring the extent of alienation. *Lifestyles*, 8(3–4), 35–46.

Dickinson, A., & Hill, R. L. (2007). Keeping in touch: talking to older people about computers and communication. *Educational Gerontology*, 33(8), 613–630.

Douglas, G., Corcoran, C., & Pavey, S. (2006). Network 1000. Opinions and circumstances of visually impaired people in Great Britain: report based on over 1,000 interviews. England: University of Birmingham.

Evjemo, B., Svendsen, G. B., Rinde, E., & Johnsen, J. K. (2004). Supporting the distributed family: the need for a conversational context. *NordiCHI'04: Proceedings of the Third Nordic Conference on Human-Computer Interaction*, pp. 309–312. New York: ACM.

Follmer, S., Raffle, H., Go, J., Ballagas, R., & Ishii, H. (2010). Video play: playful interactions in video conferencing for long-distance families with young children. *IDC'10: Proceedings of the 9th International Conference on Interaction Design and Children* (pp. 49–58). New York: ACM.

Follmer, S., Ballagas, R., Raffle, H., Spasojevic, M., & Ishii, H. (2012). People in books: Using a flashcam to become part of an interactive book for connected reading. *CSCW'12: Proceedings of the ACM Conference on Computer Supported Cooperative Work* (pp. 685–694).

Franks, L. J., Hughes, J. P., Phelps, L. H., & Williams, D. G. (1993). Intergenerational influences on Midwest college students by their grandparents and significant elders. *Educational Gerontology, 19*(3), 265–271.

Frohlich, D., Kuchinsky, A., Pering, C., Don, A., & Ariss, S. (2002). Requirements for photoware. *CSCW'02: Proceedings of the ACM Conference on Computer Supported Cooperative Work* (pp. 166–175). New York: ACM.

Gatchel, R. J., Peng, Y. B., Peters, M. L., Fuchs, P. N., & Turk, D. C. (2007). The biopsychosocial approach to chronic pain: scientific advances and future directions. *Psychological Bulletin*, 133(4), 581–624.

Gaver, W., Boucher, A., Bowers, J., Blythe, M., Jarvis, N., Cameron, D., Kerridge, T., Wilkie, A., Phillips, R., & Wright, P. (2011). The photostroller: supporting diverse care home residents in engaging with the world. *CHI'11: Proceedings of the SIGCHI Conference on Human Factors in Computing Systems* (pp. 1757–1766). New York: ACM.

Geurts, T., Tilburg, T. G., van, & Poortman, A. R. (2011). The grandparent–grandchild relationship in childhood and adulthood: a matter of continuation? *Personal Relationships*, Online First April 2011. doi:10.1111/j.1475-6811.2011.01354.x.

Glass, J., Bengtson, V. L., & Dunham, C. C. (1986). Attitude similarity in three-generation families: socialization, status inheritance, or reciprocal influence? *American Sociological Review*, 51(5), 685–698.

Hartshorne, T. S., & Manaster, G. J. (1982). The relationship with grandparents: contact, importance, role conception. *The International Journal of Aging and Human Development*, 15(3), 233–245.

Harwood, J. (2000a). Communication media use in the grandparent–grandchild relationship. *Journal of Communication, 50*(4), 56–78.

Harwood, J. (2000b). Communicative predictors of solidarity in the grandparent–grandchild relationship. *Journal of Social and Personal Relationships, 17*(6), 743–766.

Hawkey, K., Inkpen, K. M., Rockwood, K., McAllister, M., & Slonim, J. (2005). Requirements gathering with Alzheimer's patients and caregivers. *ASSETS'05: Proceedings of the 7th International ACM SIGACCESS Conference on Computers and Accessibility*, pp. 142–149. New York: ACM.

Hodgson, L. G. (1998). Grandparents and older grandchildren. In M. Szinovacz (Ed.), *Handbook on grandparenthood* (pp. 170–183). Westport: Greenwood.

Hoff, A. (2007). Patterns of intergenerational support in grandparent–grandchild and parent-child relationships in Germany. *Ageing & Society, 27*(05), 643–665.

Jamison, R. N., & Virts, K. L. (1990). The influence of family support on chronic pain. *Behaviour Research and Therapy, 28*(4), 283–287.

Judge, T. K., Neustaedter, C., & Kurtz, A. F. (2010a). Sharing conversation and sharing life: video conferencing in the home. *CHI'10: Proceedings of the SIGCHI Conference on Human Factors in Computing Systems* (pp. 655–658). New York: ACM.

Judge, T. K., Neustaedter, C., & Kurtz, A. F. (2010b). The family window: the design and evaluation of a domestic media space. *CHI'10: Proceedings of the SIGCHI Conference on Human Factors in Computing Systems* (pp. 2361–2370). New York: ACM.

Kahana, B., & Kahana, E. (1970). Grandparenthood from the perspective of the developing grandchild. *Developmental Psychology, 3*(1), 98–105.

Kemp, C. L. (2005). Dimensions of grandparent–adult grandchild relationships: from family ties to intergenerational friendships. *Canadian Journal on Aging, 24*(2), 161–177.

Keyes, C. L. M. (2002). The exchange of emotional support with age and its relationship with emotional well-being by age. *Journal of Gerontology: Psychological Sciences, 57B*(6), 518–525.

Khoo, E. T., Merritt, T., & Cheok, A. D. (2009). Designing physical and social intergenerational family entertainment. *Interacting with Computers, 21*(1–2), 76–87.

Kirk, D., Sellen, A., Rother, C., & Wood, K. (2006). Understanding photowork. *CHI'06: Proceedings of the SIGCHI Conference on Human Factors in Computing Systems* (pp. 761–770). New York: ACM.

Kirk, D., Sellen, A., & Cao, X. (2010). Home video communication: mediating closeness. *CSCW'10: Proceedings of the ACM Conference on Computer Supported Cooperative Work* (pp. 135–144). New York: ACM.

Kornhaber, A., & Woodward, K. L. (1985). *Grandparents/grandchildren: the vital connection.* New Brunswick: Transaction.

Lee, M. L., & Dey, A. K. (2007). Providing good memory cues for people with episodic memory impairment. *ASSETS'07: Proceedings of the 9th International ACM SIGACCESS Conference on Computers and Accessibility* (pp. 131–138). New York: ACM.

Lindley, S. E. (2011). Shades of lightweight: supporting cross-generational communication through home messaging. *Universal Access in the Information Society,* Online First June 2011. doi:10.1007/s10209-011-0231-2.

Lindley, S. E., Harper, R., & Sellen, A. (2009) Desiring to be in touch in a changing communications landscape: attitudes of older adults. CHI'09: Proceedings of the SIGCHI Conference on Human Factors in Computing Systems (pp. 1693–1702). New York: ACM.

Litwin, H. (1998). Social network type and health status in a national sample of elderly Israelis. *Social Science and Medicine, 46*(4–5), 599–609.

Miller, A., & Edwards, W. K. (2007). Give and take: a study of consumer photo-sharing culture and practice. *CHI'07: Proceedings of the SIGCHI Conference on Human Factors in Computing Systems* (pp. 347–356). New York: ACM.

Mills, T. L. (1999). When grandchildren grow up: role transition and family solidarity among baby boomer grandchildren and their grandparents. *Journal of Aging Studies, 13*(2), 219–239.

Mynatt, E. D., Essa, I., & Rogers, W. (2000). Increasing the opportunities for aging in place. *CUU'00: Proceedings of the Conference on Universal Usability* (pp. 65–71). New York: ACM.

Mynatt, E. D., Rowan, J., Craighill, S., & Jacobs, A. (2001). Digital family portraits: supporting peace of mind for extended family members. *CHI'01: Proceedings of the SIGCHI Conference on Human Factors in Computing Systems* (pp. 333–340). New York: ACM.

Peterson, C. C. (1999). Grandfathers' and grandmothers' satisfaction with the grandparenting role: seeking new answers to old questions. *International Journal of Aging & Human Development, 49*(1), 61–78.

Raffle, H., Ballagas, R., Revelle, G., Horii, H., Follmer, S., Go, J., Reardon, E., Mori, K., Kaye, J., & Spasojevic, M. (2010). Family story play: reading with young children (and Elmo) over a distance. *CHI'10: Proceedings of the SIGCHI Conference on Human Factors in Computing Systems* (pp. 1583–1592). New York: ACM.

Raffle, H., Revelle, G., Mori, K., Ballagas, R., Buza, K., Horii, H., Kaye, J., Cook, K., Freed, N., Go, J., & Spasojevic, M. (2011). Hello, is grandma there? Let's read! StoryVisit: family video chat and connected e-books. *CHI'11: Proceedings of the SIGCHI Conference on Human Factors in Computing Systems* (pp. 1195–1204). New York: ACM.

Robertson, J. F. (1976). Significance of grandparents: perceptions of young adult grandchildren. *The Gerontologist, 16*(2), 137–140.

Romero, N., Markopoulos, P., Baren, J., Ruyter, B., Ijsselsteijn, W., & Farshchian, B. (2007). Connecting the family with awareness systems. *Personal Ubiquitous Computing, 11*(4), 299–312.

Rowe, M., Lane, S., & Phipps, C. (2007) Carewatch: a home monitoring system for use in homes of persons with cognitive impairment. *Topics in Geriatric Rehabilitation: Smart Technology, · 23*(1), 3–8.

Ruiz, S. A., & Silverstein, M. (2007). Relationships with grandparents and the emotional well-being of late adolescent and young adult grandchildren. *Journal of Social Issues, 63*(4), 793–808.

Ryan, E. B., Anas, A. P., Hummert, M. L., & Laver-Ingram, A. (1998). Young and older adults' views of telephone talk: conversation problems and social uses. *Journal of Applied Communication Research, 26*(1), 83–98.

Seponski, D. M., & Lewis, D. C. (2009). Caring for and learning from each other: a grounded theory study of grandmothers and adult granddaughters. *Journal of Intergenerational Relationships, 7*(4), 394–410.

Sheehan, N. W., & Petrovic, K. (2008). Grandparents and their adult grandchildren: Recurring themes from the literature. *Marriage & Family Review, 44*(1), 99–124.

Smith, K. L., Crete-Nishihata, M., Damianakis, T., Baecker, R. M., & Marziali, E. (2009). Multimedia biographies: a reminiscence and social stimulus tool for persons with cognitive impairment. *Journal of Technology in Human Services, 27*(4), 287–306.

Snelgrove, W. X., & Baecker, R. M. (2010). A system for the collaborative reading of digital books with the partially sighted. *BooksOnline'10: Proceedings of the Third Workshop on Research Advances in Large Digital Book Repositories and Complementary Media* (pp. 47–50). New York: ACM.

Streltzer, A. (1979). A grandchildren's group in a home for the aged. *Health and Social Work, 4*(1), 167–183.

Tee, K., Brush, A. B., & Inkpen, K. M. (2009). Exploring communication and sharing between extended families. *International Journal of Human-Computer Studies, 67*(2), 128–138.

Tomaka, J., Thompson, S., & Palacios, R. (2006). The relation of social isolation, loneliness, and social support to disease outcomes among the elderly. *Journal of Aging and Health, 18*(3), 359–384.

Turcotte, M., & Schellenber, G. (2006). *A portrait of seniors in Canada*. Catalogue number 89-519-XIE. Ottawa: Statistics Canada.

Uhlenberg, P. (1996). Mortality decline in the twentieth century and supply of kin over the life course. *The Gerontologist, 36*(5), 681–685.

Uhlenberg, P. (2004). Historical forces shaping grandparent–grandchild relationships: demography and beyond. *Annual Review of Gerontology and Geriatrics, 24*, 77–97.

Vetere, F., Davis, H., Gibbs, M., & Howard, S. (2009) The magic box and collage: responding to the challenge of distributed intergenerational play. *International Journal of Human-Computer Studies, 67*(2), 165–178.

Voida, A., & Greenberg, S. (2009) Wii all play: the console game as a computational meeting place. *CHI'09: Proceedings of the SIGCHI Conference on Human Factors in Computing Systems* (pp. 1559–1568). New York: ACM.

Vutborg, R., Kjeldskov, J., Pedell, S., & Vetere, F. (2010). Family storytelling for grandparents and grandchildren living apart. *NordiCHI'10: Proceedings of the 6th Nordic Conference on Human-Computer Interaction* (pp. 531–540). Reykjavik, Iceland.

Wu, M., Baecker, R., & Richards, B. (2007). Designing a cognitive aid for and with people who have anterograde amnesia. In J. Lazar (Ed.), *Universal usability* (pp. 317–356). West Sussex: Wiley.

Index

C. Neustaedter et al. (eds.), *Connecting Families,*
DOI 10.1007/978-1-4471-4192-1, © Springer-Verlag London 2013